TOOL STEELS
Properties and Performance

TOOL STEELS
Properties and Performance

Rafael A. Mesquita

CRC Press
Taylor & Francis Group
Boca Raton London New York

CRC Press is an imprint of the
Taylor & Francis Group, an **informa** business

CRC Press
Taylor & Francis Group
6000 Broken Sound Parkway NW, Suite 300
Boca Raton, FL 33487-2742

First issued in paperback 2020

© 2017 by Taylor & Francis Group, LLC
CRC Press is an imprint of Taylor & Francis Group, an Informa business

No claim to original U.S. Government works

ISBN-13: 978-1-4398-8171-2 (hbk)
ISBN-13: 978-0-367-78257-3 (pbk)

Visit the Taylor & Francis Web site at
http://www.taylorandfrancis.com

and the CRC Press Web site at
http://www.crcpress.com

Contents

Preface

Seeing the final format of this book gives me a pleasant feeling of accomplishment and, at the same time, of gratitude: first, because it comes after more than 15 years of a career with extensive dedication to tool steels, and second, because I can see in the upcoming chapters many concepts that I learned from my university teachers, with other researchers in various fields, many of whom I have had the opportunity to meet in person, and also from coworkers and friends in the field dealing with applications of tool steels. This book also brings me nostalgia, since tool steels is the topic that I started my career with and, still today, I am impressed with the rich metallurgical content in this amazing class of materials.

Another challenging aspect is the responsibility of writing a book that is worthy to follow the many excellent references on tool steels, which is not that simple: for example, the famous *Tool Steels*, with numerous editions since 1944 by G. Roberts and coauthors and the last edition in 1998 by Roberts, Krauss, and Kennedy. Having in this last edition George Krauss, one of the most important references in steel physical metallurgy, is of special remark for me; I have met him and we have exchanged ideas, on several occasions, on heat treatment and other general metallurgical concepts in steels. The present book also brings me the responsibility of replicating the knowledge that I gained from numerous teachers, with special recognition to my PhD advisor Professor H.-J. Kestenbach, who first introduced me to "think steel metallurgy" based on microstructure. Finally, it is also a big responsibility to explain the industrial aspect of tool steel technologies and applications using published results from tool steel producers who are referenced in industry, mainly from the Voestalpine Edelstahl group, namely, Villares Metals, Uddeholm Tooling, Böhler Edelstahl, and Buderus Edelstahl—companies that I worked with and represent the benchmark of applied knowledge of tool steels in the industry.

After considering all those aspects of tool steels and my professional life, starting the book was not simple. On one hand, so many ideas were flowing, most of them motivated by my passion for this topic. On the other hand, putting them in order and selecting what should be in the book and what should be left aside was also a challenge, in order to finish the book with a reasonable size and in proper time. I then followed the suggestion of a great book writer and personal friend, Dr. George Totten, who once told me, "Think about the graphs, tables, pictures and other information that you are constantly needing to check or refer to in your professional life. Those are the important topics for your book." So, this book contains most of the things I used in my career, both in academia and industry. In fact, the invitation for the book started when I was in academia, after a long period in industry, and the book was finished when I returned to industry. The text thus blends concepts with a solid background of physical metallurgy and observations and remarks on typical industrial application, and it can be used in both fields.

The book starts with an introduction in Chapter 1 to tool steels and its definitions, as well as a small discussion on the designations for different types of tool steels in different standards and different countries. Following this introduction, Chapter 2 brings the state of the art in melt shop technologies used to appropriately produce a tool steel, with controlled chemical composition, but also considering all the purity aspects, in terms of presence of nonmetallic inclusions. The chapter also presents the forming aspects and how they interact with the development of the microstructure, which in turn will determine the properties. Authors Kubin Michael and Reinhold Schneider, who have many years of experience in the industrial production of tool steels, kindly offered their contribution to the book and wrote this chapter.

Chapter 3 then deeply explains this relation between microstructure and properties. It contains the physical metallurgy and the heat treatment concepts for tool steels. It is likely that readers may skip and go directly to what matters: the manufacturing (Chapter 2) or the applications of specific types of tool steels (Chapters 4 through 7). However, Chapter 3 is extremely important and I would strongly suggest the opposite, using Chapter 3 as a basis for all of the other concepts. Tool steels constitute a complex class of materials, and Chapter 3 breaks this complexity by separating their microstructure into two major fields: the microstructural matrix, including the modifications caused by heat treatment, and the particles, describing the importance of carbide type, size, and distribution. A common basis is then proposed for all types of tool steels, namely "microstructural approach," applicable to the most diverse classes of tool steels, from mold steels to high-speed steels.

From the basis given in Chapters 1 through 3, the following chapters may be understood. Chapter 4 shows the property requirements and the main aspects of the metallurgy of hot work steels, often bringing the main applications of extrusion, die-casting, and forging. The two main important properties for those steels, hot strength and toughness, are the focus of this chapter. Tempering relationship with hot strength is presented, as well as the development of manufacturing and heat treatment to promote high toughness. Nitriding is extensively applied in hot work steels, and thus a short discussion on this surface treatment is presented. As in other chapters, examples of failures are used to illustrate the importance of the developed concepts.

Chapter 5 is dedicated to cold work tool steels, discussing the main needs of wear resistance and the properties related to the abrasive and adhesive wear mechanisms. Hardness, toughness, and the distribution of carbides are the main features of Chapter 5, as well as the interaction of manufacturing process and heat treatment to the final microstructure and properties. Due to the usual high volume in cold work steel microstructure, retained austenite is discussed in depth in this chapter which, together with Chapter 3, may be used to give the basis for understanding the effects of this phase in other tool steel types as well.

In Chapter 6, plastic mold steels are presented. In terms of microstructure, they contain similar characteristics to hot work steels, but they have other requirements, such as achieving high surface quality for polishing or texturing and also achieving good through-hardening with minimal cost. Both aspects are directly dependent on the manufacturing characteristics of the plastic mold steels, such as the machining, polishing, texture etching, and the quenching. The dual approach of the manufacturing conditions and the final properties is then a constant topic in this chapter, always using the microstructure to link both.

Finally, high-speed steels are approached in Chapter 7. This is a very complex group of tool steels with perhaps the largest amount of information available in literature. However, cutting tools, which today represent the main application of high-speed steels, also use other groups of materials such as solid carbides or, in smaller extent, ceramic-based compounds. Therefore, a general comparison on materials for cutting tools is first presented, followed by a discussion on high-speed steels applied in cutting tools in terms of hardness, heat treatment, carbide type, and as-cast carbide distributions. The possible application of high-speed steels in forming tools for hot work or cold work is mentioned in some points, but is also explained in their respective Chapters 4 and 5.

As a last comment, all specific chapters, Chapters 4 through 7, present at the end a semiquantitate comparison of properties for different grades. This may be useful for fast decisions and mainly to understand the general characteristics of the most common grades in one single picture. A brief discussion of the main important grades was also written after this comparison, with more focus to the reference grades for each class: H13, D2, mod. P20 (DIN 1.2738), and M2.

I would like to acknowledge the great efforts from Eng. MSc. Paulo T. R. Haddad who, with his extensive experience and knowledge in tool steels, revised all chapters and gave excellent suggestions to the improvement of the final text of this book.

It is my personal belief that this book substantially contributes to update and summarize the important aspects in the properties and performance of tool steels.

Have a good reading!

Dr. Rafael A. Mesquita
Ann Arbor, Michigan

1 Introduction to Tool Steels

Various aspects of tool steels are presented and discussed in this book. This chapter will first guide the reader on the main attributes of tool steels, how this class of materials may be segmented, and how the main metallurgical aspects are interconnected. Starting with the definition of tool steels, this chapter then introduces the four main groups of tool steels, which will form the base for the discussion in the subsequent chapters. And this division is based on the final tooling application, rather than on composition or other metallurgical aspects.

Therefore, despite addressing the properties, microstructure, and manufacturing in depth, this book will be driven by understanding the performance on each application of the especially very interesting class of ferrous materials called *tool steels*. In this introductory chapter, tool steels are defined, the main standards used for them are described, and the classification of tool steels is presented, which will be the same classification followed in the next chapters.

1.1 DEFINITION OF TOOL STEELS

Before starting to write on any topic, the first attempt is to try to find a precise definition for it in order to establish the rationale for separating it from other areas. This will be the purpose of this chapter, which gives the initial directions on tool steels.

A precise definition of tool steels, however, is not found in the literature. Tool steels do not correspond to a specific chemical composition or have a specific set of properties or applications. One of the best definitions found in the literature is given in the excellent book of Roberts et al. [1], when citing the description of "Tool Steels" in the *Steel Products Manual of Iron and Steel Society*:

> Tool steels are (1) carbon-, alloy-, or high-speed steels, capable of being hardened and tempered. They are usually melted in electric furnaces and (2) produced under tool steel practice to meet special requirements. They may be used in certain hand tools or in mechanical fixtures for (3) cutting, shaping, forming, and blanking of materials at either ordinary or elevated temperatures. Tool steels are also used in a wide variety of other applications where resistance to wear, strength, toughness, and other properties are selected for optimum performance.

This description is composed of three major elements, which are numbered according to the transcription from Iron Steel Society text, reference [2], and summarized in Figure 1.1. The first element of the above definition points to the fact that tool steels may show a wide range of chemical compositions, from very low alloy grades

1

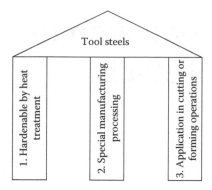

FIGURE 1.1 Schematic representation of the main pillars that define tool steels.

to compositions with extremely high amounts of alloy elements. But one characteristic is common to all these groups: their ability to enhance the properties by means of heat treatment, especially by hardening and tempering. This characteristic is very remarkable indeed in tool steels, as it represents a tremendous enhancement in their properties after hardening and tempering. For example, tool steels are perhaps the only class of materials in nature that are able to cut themselves: a hardened high-speed steel tool is capable of drilling bars also made of high-speed steel when those are in the untreated or, more precisely speaking, annealed form. This potential provided by the heat treatment of tool steels also makes the latter dependent on the application of proper heat-treatment conditions, which also depend on their metallurgical aspect. In this sense, Chapter 3 will deal with all the metallurgical features of different tool steels and include an in-depth discussion on their heat-treatment aspect.

The second part of the definition of tool steels concerns the manufacturing methods used to produce a tool steel in the steel mill. When compared to the amount of carbon and low-alloy steels produced, which is in the range of two billion tons per year, tool steels are a very small group and represent about 0.1% of that (probably around two million tons per year, worldwide, depending on which grades are classified as tool steels). This lower volume produced means that tool steel production methods may be different from those of large-volume steels due to simple scale factors. In addition, the superior properties required of tool steels creates the need for careful control of their process and, in most cases, addition of manufacturing operations that are not common in other steels. So, these manufacturing aspects of tool steels also differentiate them from other steels, constituting the second pillar in the definition of tool steels detailed in Chapter 2.

The last aspect in defining a tool steel is the final application: used for shaping, forming, or cutting ferrous alloys, other metallic alloys, and other nonmetallic materials, at room temperature or high temperatures (usually up to 1200°C). The range of applications of tool steels is therefore very wide and also large while combining the steel grades, heat treatment conditions, and manufacturing methods, in addition to tool mechanical design. This characteristic is probably the most important topic in this book and is used to group different types of tool steels within groups for a given application: cold work tool steels, hot work tool steels, plastic mold tool steels, and high-speed cutting tool steels, the last group being well known only for high-speed steels. These aspects are then described in the next section of this introductory chapter and are used to describe the specific characteristics of each group in later chapters (Chapters 4 through 7).

1.2 DESIGNATION OF TOOL STEELS

There are many different designations for tool steels, based on different standards in different countries (e.g., United States, China, Europe/Germany, Japan) and also commercial designations based on proprietary compositions, or simply different names for marketing purposes. Therefore, it is never easy to define precisely the tool steel by looking at its name; nor is it easy to compare steels from different standards and different manufactures. Some books or encyclopedias may help in that respect, gathering most of the grades and their typical composition, such as the Stahlschluessel, which means "key to steels" (see Reference 3); although it is a German title, it has an English text content.

One of the main classification methods for tool steels is the ASTM (American Society for Testing and Materials) designation [4,5], where each steel grade is named by a letter followed by a number. The letter means one characteristic of the tool steel, and the number, usually in sequential (and usually historical) order, denotes a given developed steel. Table 1.1 shows all letter designations in the ASTM group of tool steels. For example, the tool steel A2 means that it belongs to the "Air-hardenable group" of tool steels. In steel H13, however, the letter "H" refers to its application, namely in hot-forming processes. And, as a last example, in the steel M2, the letter M refers to the element molybdenum, which is the main alloying element providing the properties of hot hardness and wear resistance in cutting tools and thus having a chemical composition relation. So it is clear that there is no single parameter for grouping one or another steel into a specific class within the ASTM classification. In addition to this classification, steels are also identified by designations in the Unified

TABLE 1.1
ASTM Classification of Tool Steels

Classification ASTM	Symbol
Water-hardening tool steels	W
Shock-resisting tool steels	S
Oil-hardening cold work tool steels	O
Air-hardening, medium-alloy cold work tool steels	A
High-carbon, high-chromium cold work tool steels for Dies	D
Plastic mold steels	P
Hot work tool steels, chromium, tungsten	H
Tungsten high-speed tool steels	T
Molybdenum high-speed tool steels	M

Numbering System (UNS) for Metals and Alloys, established in 1975 by ASTM and Society of Automotive Engineers (SAE).

Another important standard tool steel classification is the one based on the former DIN 17350 Standard [6], currently the ISO Standard number 4957 [7]. This classification gives a designation based on the chemical composition and also a number for each grade. For example, the steel H13 in the EN is named X40CrMoV5-1 or by the number 1.2344. The first letter is a reference to the main chemical elements, the letter X followed by 100 times the carbon content, so when reading X40CrMoV5-1, one knows that this is a 0.40C steel alloyed with Cr, Mo, and V. The main grouping is based on a single criterion, the final application, thus dividing the steels into unalloyed or alloyed cold work tool steels, hot work tool steels, or high-speed steels. Plastic molds are included in cold work steels, which makes sense because the process temperatures are not so high as to affect the steel microstructure such as what happens in hot work or high-speed steels.

Other standards also exist, from different countries: Japan (JIS), the United Kingdom (BS), France (AFNOR), or Sweden (SSH). A cross-reference of the main grades may be built, which will not be exact since the composition ranges may vary between the various standards. But this cross-reference is always useful, as is presented in Table 1.2.

But those are much less common today, and most literature refers to the ASTM (former AISI) standard, the EN (former DIN), and especially the 1.XXXX numbers, or the commercial designation of steels developed by companies which are not standardized. This will then be used in this book, which will focus more on the ASTM designation and less on the DIN designation.* Commercial designations of steel grades that have not been standardized are avoided, but in some cases they are necessary and also used.

1.3   CLASSIFICATION OF TOOL STEELS

As in any area of scientific knowledge, grouping similar tool steels is a useful method to facilitate their understanding. It is clear from the aforementioned examples that the ASTM designation does not show the common elements for classifying tool steels, since the criteria change from heat treatment, chemical composition, or application. In terms of EN standard, the classification based only on application is more coherent because it deals only with the final application. However, plastic mold steels are classified into cold work steels, but there is a division based on chemical composition in cold work steels themselves.

The classification proposed in this book follows that predominantly used in industry, that is, close to the EN but purely based on the final application. Four main groups may then be identified: hot work tool steels, cold work tool steels, plastic mold steels, and high-speed steels. This way of segmenting tool steels is then used

---

* Although the EN is the current designation, the numbers are the same as the former DIN numbers and, as the industry is more familiar with the DIN numbers, to keep the practical aspect of the present book, the DIN designation is still referred to instead of EN.

TABLE 1.2

Cross-Reference of the Main Tool Steels, according to Different Standards for Tool Steel Compositions and Designations

United States (AISI)	West Germany (DIN)(a)	Japan (JIS)(b)	Great Britain (B.S.)(c)	France (AFNOR)(d)	Sweden (SS14)
Molybdenum high-speed steels (ASTM A 600)					
M1	1.3346	—	4659 BMI	A35-590 4441 Z85 DCWV08-04-02-01	2715
M2, reg C	1.3341, 1.3343, 1.3345, 1.3553, 1.3554	G4403 SKH51 (SKH9)	4659 BM2	A35-590 4301 Z85WDCV06-05-04-02	2722
M2, high C	1.3340, 1.3342	—	—	A35-590 4302 Z90WDCV06-05-04-02	—
M3, class 1	—	G4403 SKH52	—	—	—
M3, class 2	1.3344	G4403 SKH53	—	A35-590 4360 Z120 WDCV06-05-04-03	(USA M3 class 2)
M4	—	G4403 SKH54	4659 BM4	A35-590 4361 Z130 WDCV06-05-04-04	—
M7	1.3348	G4403 SKH58	—	A35-590 4442 Z100DCWV09-04-02-02	2782
M35	1.3243	G4403 SKH55	—	A35-590 4371 Z85WDKCV06-05-05-04-02	—
				A35-590 4372 Z90WDKCV06-05-05-04-02	
M42	1.3247	G4403 SKH59	4659 BM42	A35-590 4475 Z110DKC WV09-08-04-02	—
Intermediate high-speed steels					
M50	1.2369, 1.3551	—	—	A35-590 3551 Y80DCV42.16	(USA M50)
M52	—	—	—	—	—
Tungsten high-speed steels (ASTM A 600)					
T1	1.3355, 1.3558	G4403 SKH2	4659 BT1	A35-590 4201 Z80WCV18-04-01	—
T2	—	—	4659 BT2, 4659 BT20	4203 18-0-2	—
T4	1.3255	G4403 SKH3	4659 BT4	A35-590 4271 Z80WKCV18-05-04-01	—
T5	1.3265	G4403 SKH4	4659 BT5	A35-590 4275 Z80WKCV18-10-04-02	(USA T5)
T15	1.3202	G4403 SKH10	4659 BT15	A35-590 4171 Z160WKVC12-05-05-04	(USAT15)

(Continued)

TABLE 1.2 (Continued)
Cross-Reference of the Main Tool Steels, according to Different Standards for Tool Steel Compositions and Designations

United States (AISI)	West Germany (DIN)(a)	Japan (JIS)(b)	Great Britain (B.S.)(c)	France (AFNOR)(d)	Sweden (SS14)
Chromium hot work steels (ASTM A 681)					
H10	1.2365, 1.2367	G4404 SKD7	4659 BH10	A35-590 3451 32DCV28	—
H11	1.2343, 1.7783, 1.7784	G4404 SKD6	4659 BH11	A35-590 3431 FZ38CDV5	—
H12	1.2606	G4404 SKD62	4659 BH 12	A35-590 3432 Z35CWDV5	—
H13	1.2344	G4404 SKD61	4659 BH13, 4659 H13	A35-590 3433 Z40CDV5	2242
H14	1.2567	G4404 SKD4	—	3541 Z40WCV5	—
H19	1.2678	G4404 SKD8	4659 BH19	—	—
Tungsten hot work steels (ASTM A 681)					
H21	1.2581	G4404 SKD5	4659 BH21, 4659 H21A	A35-590 3543 Z30WCV9	2730
Molybdenum hot work steels (ASTM A 681)					
H42	—	—	—	3548 Z65WDCV6.05	—
Air-hardening, medium-alloy, cold work steels (ASTM A 681)					
A2	1.2363	G4404 SKD12	4659 BA2	A35-590 2231 Z100CDV5	2260
A7	—	—	—	—	—
A8	1.2606	G4404 SKD62	—	3432Z38CDWV5	—
A9	—	—	—	—	—
A10	—	—	—	—	—

(Continued)

TABLE 1.2 (Continued)
Cross-Reference of the Main Tool Steels, according to Different Standards for Tool Steel Compositions and Designations

United States (AISI)	West Germany (DIN)(a)	Japan (JIS)(b)	Great Britain (B.S.)(c)	France (AFNOR)(d)	Sweden (SS14)
High-carbon, high-chromium cold work steels (ASTM A 681)					
D2	1.2201, 1.2379, 1.2601	G4404 SKD11	4659 (USA D2), 4659 BD2 4659 BD2A	A35-590 2235 Z160CDV12	2310
D3	1.2080, 1.2436, 1.2884	G4404 SKD1, G4404 SKD2	4659 BD3	A35-590 2233 Z200C12	—
D4	1.2436, 1.2884	G4404 SKD2	4659 (USA D4)	A35-590 2234 Z200CD12	2312
D5	1.2880	—	—	A35-590 2236 ZI 60CKDV 12.03	—
D7	1.2378	—	—	2237 Z230CVA12.04	—
Oil-hardening cold work steels (ASTM A 681)					
O1	1.2510	G440-4 SKS21, G4404 SKS3, G4404 SKS93 G4404 SKS94, G4404 SKS95	4659 BO1	A35-590 2212 90 MWCV5	2140
O2	1.2842	—	4659 (USA 02) 4659 BO2	A35-590 2211 90MV8	—
O6	1.2206	—	—	A35-590 2132 130C3	—
O7	1.2414, 1.2419, 1.2442, 1.2516, 1.2519	G4404 SKS2	—	A35-590 2141 105WC13	—
Shock-resisting steels (ASTM A 681)					
S1	1.2542, 1.2550	G4404 SKS41	4659 BS1	A35-590 2341 55WC20	2710
S2	1.2103	—	4659 BS2	A35-590 2324 Y45SCD6	—

(Continued)

TABLE 1.2 (Continued)
Cross-Reference of the Main Tool Steels, according to Different Standards for Tool Steel Compositions and Designations

United States (AISI)	West Germany (DIN)(a)	Japan (JIS)(b)	Great Britain (B.S.)(c)	France (AFNOR)(d)	Sweden (SSI4)
S5	1.2823	—	4659 BS5	—	—
S6	—	—	—	—	—
S7	—	—	—	—	—
Low-alloy special-purpose steels (ASTM A 681)					
L2	1.2235, .1.2241, 1.2242, 1.2243,	G4404 SKT3, G4410 SKC11	—	A35-590 3335 55CNDV4	—
L6	1.2713, 1.2714	G4404 SKS51, G4404 SKT4	—	A35-590 3381 55NCDV7	—
Low-carbon mold steels (ASTM A 681)					
P20	1.2738, 1.2311, 1.2328, 1.2330	—	4659 (USA P20)	A35-590 2333 35CMD7	(USAP20)
Water-hardening steels (ASTM A 686)					
W1	1.1525, 1.1545, 1.1625, 1.1654, 1.1663, 1.1673, 1.1744, 1.1750, 1.1820, 1.1830	G4401 SKI, G4401 SK2, G4401 SK3, G4401 SK4,G4401 SK5, G4401 SK6 G4401 SK7, G4401 SK7,G4410SKC3	4659 (USA WI), 4659 BW 1A 4659 BW1B 4659 B,W1C	A35-590 1102 Y(I) 105, A35-5901103 Y(1)90 A35-590 1104 Y(1) 80, A35-590 1105 Y(1) 70 A35-590 1200 Y(2) 140, A35-5901201 Y(2) 120 A35-596 Y75, A35-596 Y90	—

(Continued)

TABLE 1.2 (Continued)

Cross-Reference of the Main Tool Steels, according to Different Standards for Tool Steel Compositions and Designations

United States (AISI)	West Germany (DIN)(a)	Japan (JIS)(b)	Great Britain (B.S.)(c)	France (AFNOR)(d)	Sweden (SS14)
W2	1.1645, 1.2206, 1.283	G4404 SKS43 G4404 SKS44	4659 BW2	A35-590 1161 Y120V, A35-590 1162 Y105V A35-590 1163 Y90V, A35-590 1164 Y75V	(USA W2A) (USA W2B) (USA W2C)
W5	1.2002, 1.2004, 1.2056	—	—	A35-590 1230 Y(2) 140C, A35-590 2130 Y100C2 A35-590 1232 Y105C	—

Sources: *Stahlschluessel*, 23. Auflage. Edition, 2013, n(January 1, 2013); ASTM A681-08, Standard specification for tool steels alloy, ASTM International, Materials Park, OH.

Notes: (a) Deutsche Industrie Normen (German Industrial Standards), (b) Japanese Industrial Standard, (c) British Standard. (d) l'Association Française de Normalisation (French Standards Association); Some of those standards are not applicable anymore, but they are still referred to in industrial literature and this cross-reference is always useful.

TABLE 1.3

Classification of Tool Steels Used in This Book, Including the Chapters Dedicated to Each, the Main Properties, the ASTM Classes inside Each Group, and the Main Representative Grades of Each Group

Classification in This Book	Main Properties	ASTM Classes Involved	Main Representative ASTM/DIN
Hot Work Tool Steels, Chapter 4	Hot strength Toughness	H	H13/1.2344
Cold Work Tool Steels, Chapter 5	Hardness	W, S, O, A, D	D2/1.2379
Plastic Mold Steels, Chapter 6	Hardness and hardenability Surface finishing Machinability	P	Mod. P20/1.2738
High-Speed Steels, Chapter 7	Hot hardness Wear resistance	M, T	M2/1.3343

throughout this book, and each specific class of tool steels is presented and discussed in Chapters 4–7, respectively, after a clear definition of the manufacturing conditions in Chapter 2 and the basis of physical metallurgy, microstructure, and heat treatments (Chapter 3).

Finally, Table 1.3 shows the major characteristics of each group, as well as the correlation to the well-known classes of ASTM. This reinforces the division of topics present in this book and why those four groups of steel are the most accepted and used in the industry.

REFERENCES

1. G. Roberts, G. Krauss, R. Kennedy, *Tool Steels*, 5th edn. Materials Park, OH: ASM International, 1998, pp. 7–28.
2. *Steel Products Manual, Section Tool Steels* [s.l]. ISS, Iron and Steel Society, 1988, 81pp.
3. *Stahlschluessel*, 23. Auflage. Edition, 2013, n(January 1, 2013).
4. ASTM A681-08. Standard specification for tool steels alloy. Materials Park, OH: ASTM International.
5. ASTM A600-92. Standard specification for tool steel high speed. Materials Park, OH: ASTM International.
6. DIN 13350. *Tool Steels*, Deutsches Institut für Normung. 1980. Substituted by ÖNORM EN ISO 4957:1999.
7. ÖNORM EN ISO 4957:1999. *Tool Steels*. Publisher and printing: Österreichisches Normungsinstitut, 1020 Wien.

2 Manufacturing of Tool Steels

Contributing Authors: Michael Kubin and Reinhold Schneider

Knowing the types of tool steels that will be covered and their main applications in industrial tools would help us understand the metallurgy and properties of those tools. While this topic is treated extensively throughout this book, the present chapter is placed first to show the manufacturing parameters and conditions of tool steels and also to better relate to the tool steel definition seen in Chapter 1, where the production under special conditions is one of the main aspects of tool steel characteristics.

This chapter then presents the liquid-metal processing, in terms of melt-shop processes, forming, and the heat treatment operations performed during the production of tool steels in the mill. Along these three sections, examples are used concerning the effect of the process variables on the tool steel properties, supporting the explanations of each type of tool steels later presented from Chapters 3 through 7.

2.1 LIQUID PROCESSING AND CASTING

In general, the chemical composition of tool steels is characterized by a high content of various elements: carbon between 0.4% and 2.1% (powder metallurgy, or PM steels, with up to 3%C), chromium up to 12%, molybdenum up to 12%, tungsten up to 18%, vanadium up to 4% (PM up to 15%), and niobium in some cases up to 2%. The content of sulfur (<0.03%) and phosphorus (<0.03%) and, in general, a maximum content of tramp elements are often very low, with the exception of some grades for plastic molding, where high S is added to improve the machining characteristics (see Chapter 6).

All the control of the chemical composition and also the removal of undesired constituents are done in liquid processing. The achievement of the specified range of all elements is very complex, and so is the control or the removal of residual elements that are deleterious to the mechanical or other properties of tool steels. In addition to the steel itself, other phases are present in liquid steel. Those phases are often oxides, sulfides, or any other nonmetallic phases, formed by the liquid reaction with the atmosphere or as a result of elements from the raw material, and they are named nonmetallic inclusions and are usually harmful to the properties of the steel. The control and removal of those particles are then also part of the liquid processing.

The present section is divided in three main groups: (1) melting, including the conversion of steel scrap into liquid steel and the initial addition of alloying elements; (2) secondary refining, where the final adjustments in composition are made and the inclusions may be further removed or modified; and (3) casting, where the liquid steel is poured into molds and solidified into the first solid product, named ingot.

2.1.1 MELTING

2.1.1.1 General Aspects of Electric Arc Furnace

After the implementation of the ladle furnace in secondary steel making some decades ago, the electric arc furnace (EAF) is today the main—or practically the only—melting facility for the start of liquid processing of tool steels. EAF is simply a melting machine, leaving all the metallurgical aspects to the downstream refining facilities, such as the ladle furnace and the vacuum degassing discussed later. Nowadays, the common understanding is that the task of EAF is to melt all kinds of iron-bearing material, such as iron scrap, direct reduction iron (DRI), hot briquetted iron (HBI), and pig iron, using a standardized metallurgical practice, taking care of the maximum productivity and/or the minimum transformation costs. EAF is composed basically of three electrodes that melt the raw material, as shown in Figure 2.1.

From the metallurgical point of view, EAF melting consists of an oxidizing step, oxidizing carbon and other elements such as silicon or manganese, and providing at the tapping a nearly constant chemical composition. In general, the carbon content of the liquid melt is ~0.08%, and in addition to the required alloying elements, residual elements such as copper, antimony, arsenic, sulfur, and phosphorus are present.

All these statements are valid for the major part of the common steel grades with a low content of ferroalloys. When the produced steel grades contain a high content of ferroalloys, such as high-alloy tool steels or stainless steels, the EAF practice is also fully involved in metallurgical considerations. EAF metallurgy is oriented toward both clean steel production and cost reduction, avoiding losses of expensive alloying elements such as chromium and also molybdenum, vanadium, and tungsten. Chromium reacts very easily with oxygen, and this implies the risk of chromium losses, especially when a considerable amount of oxygen must be injected; in these cases, the metallic yield decreases because a considerable amount of the oxides is contained in the slag.

2.1.1.2 Melting Procedure for Tool Steels

The melting of steel grades has to follow a "clean steel route," starting from the selection of a proper raw material, high-quality recyclable scrap having a chemical composition that is similar to the end product, and a minimum percentage of undesired elements. With this focus, it is very important to avoid contamination of the scrap, especially from metals that cannot be oxidized such as Cu, Ni, and Sn. In EAF, depending on the chosen melting practice, ferroalloys may be added. EAF melting practice for tool steels is not the same as for the other commercial low-alloyed steel grades.

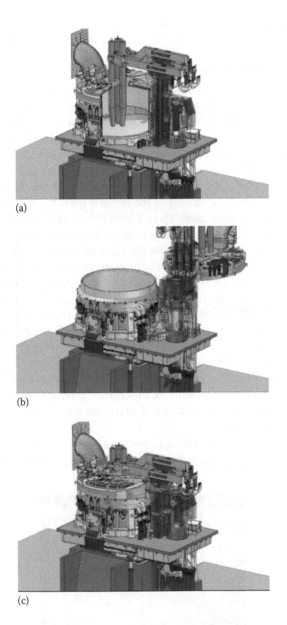

FIGURE 2.1 Example of operation of an electric arc furnace (EAF), showing (a) a cross section, with the three electrodes used to melt the scrap and other sources of metal, (b) the furnace with the lid open, enabling the addition of scrap, and (c) lid closed and the melting process going on.

There are several factors that influence the melting practice:

- The chemical composition of the raw material input
- The final content of expensive alloys such as Cr, Mo, and W
- The raw material density

To achieve high percentages of chrome in the steel, ferroalloys containing chromium in combination with high chrome-containing scrap, mainly heavy pieces of return scrap from ingot-casting processes, are used. This kind of scrap allows the recovering of a large quantity of important elements, but has the drawback of being mainly composed of big pieces, with the consequent uneven bucket charging and difficult melting conditions. The raw material density is an important factor for the management of the power input, melting rate, and utilization of the burners and oxygen injection tools. In general, the density of the scrap is quite high, reaching values well above 0.7 ton/m^3 and in some cases up to 2 tons/m^3, in addition to the big blocks.

The arc power must be adjusted to obtain sufficient arc stability, decreasing the working power factor and keeping stable the electrode regulator. The use of long arc operation is limited, and an electric arc with more current utilization is preferred.

In addition, the burners and the oxygen injectors, when used, should be installed near the steel and with a steep injection angle (see Figure 2.2). Particular care has to be taken to avoid a big piece of scrap in front of the injection point, which would cause flame flashback due to reflection. A flame flashback is very dangerous for the life of the shell's water-cooled elements and, even when lasting for only a couple of seconds, this flashback can damage the panel and create water leakage. A special feature implemented in some commercially available injectors avoids or limits this possible risk. There are also systems that monitor the temperature and thus foresee possible problems, through the analysis of increments in temperature in the copper injector box detection point.

The scrap melting rate evaluation is also important, since the melting of a big piece is mainly due to convection heat transfer with the liquid steel. Thus, it can be

FIGURE 2.2   Installation of a commercially available injector "near steel."

said that the power input density in those cases is lower than that of the commercial grades produced by traditional steel melting practice.

Another aspect to be monitored is the utilization of oxygen during the melting process. It is strongly related to the final carbon content, the presence of chromium in the charge, and the eventual necessity of dephosphorization of the liquid steel. The use of an oxidizing atmosphere is needed in case of dephosphorization treatment. Conversely, if a high amount of chromium should be kept in the liquid steel, the amount of oxygen within the process has to be kept at a very low level. This is the main reason the used scrap has to be low in phosphorus content in some high-Cr, low-P tool steels, in order to not apply dephosphorization treatment and consequently avoid Cr losses by oxidation.

2.1.1.3 Metallurgical Aspects: Slag Metallurgy in EAF

The slag management in tool steels differs from the common EAF practice applied in carbon steels or low-alloyed steels, in relation to the foaming slag. Although many stainless steel plants have tried to adopt this kind of practice, the development of foaming slag for those special steel grades has often been difficult.

The possibility of foaming the slag is related to some factors such as the basicity index (CaO/SiO_2 in the range of 2% ± 10%) and its content of FeO. When melting medium and high Cr grades, the composition and viscosity of the slag are not adequate for foaming. In Figure 2.3, it can be seen that a high amount of chrome generates a slag that is not in the optimum region. The reason for this is that the slag, generated during the production of these special grades, has a high chromium oxide content, due to the high affinity of chrome with oxygen, and, consequently, a lower iron oxide content. As a result, those slags are characterized by different properties

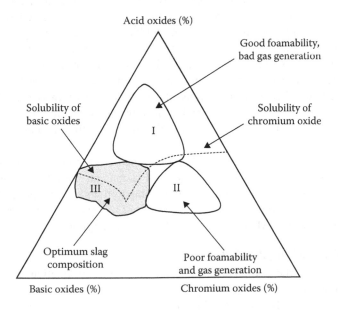

FIGURE 2.3 Qualitative ternary diagram for proper foaminess of the slag.

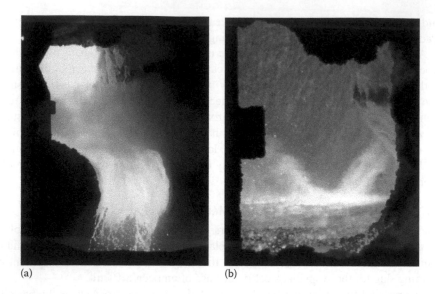

(a)   (b)

FIGURE 2.4   Images showing different types of slags: (a) fluid slag, and (b) hard slag.

than "standard" slags, in terms of hardness, viscosity, and, finally, the melting point. In Figure 2.4, the images of fluid and hard slags are shown.

It is evident that if those three slag constituents are not properly set, a hard slag will be formed, with high viscosity and low melting point, leading to low foaming capacity and low efficiency of the injection. In EAF slags, the chromium oxides have a lower solubility ($Cr_2O_3 < {\sim}5\%$) than iron oxides at the same basicity and temperature conditions. When the solubility of Cr_2O_3 is exceeded in the slag, the resulting solid phases break the foam bubbles, leading to precipitation of solid particles in the bulk and lower gas generation during foaming. This has a negative impact on the ability of the slag to foam. The optimum range for the slag composition is relatively narrow in the case of slags containing Cr_xO_y. So, achieving a stable foam in the slag is difficult, and there is a large possibility of uncontrolled oxidation with high chromium losses to slag and poor foaming behavior. Al and FeSi can be used to recover as much of the Cr as possible by a reduction reaction of the Cr_2O_3 content of the slag.

Another practice in tool steel production is the partial recarburization of the liquid steel. It is not so much related to the final C content, since the efficiency of the recarburization in the EAF is not very high compared with the tapping addition, but it is mainly related to the reduction of the ppm of O_2 in the steel for avoiding violent reactions both in the EAF and mainly in the ladle at tapping.

2.1.1.4   EAF Melting Practice

In order to cover the complete range of the tool steel production, three different EAF melting practices are generally considered, depending on the composition of the metallic charge, the content of impurities, and the final steel composition. Those three melting practices are briefly described next.

2.1.1.4.1 Standard Heats

The EAF charge is composed of carbon and alloyed steel scrap with the limitation of a maximum Cr content ≤1.0%. As discussed earlier, higher contents of chromium may produce a hard and dense slag with limited reactivity and foaming capacity. Limitation of Cr content in the charge allows the EAF operation with long arcs, foaming slag, and high input of chemical energy, ensuring the highest furnace efficiency at the lowest cost. If necessary, required alloys that do not easily react with oxygen can be added into the EAF charge.

2.1.1.4.2 Build-Up Heats

The melting process starts with a charge built up with carbon steel scrap and alloy steel scrap with the limitation of maximum chromium content ≤0.5% after meltdown. If necessary, the charge can contain added FeMo, Ni, and W. By the end of melting, liquid steel is dephosphorized and decarburized using oxygen, iron oxides, or iron ore and lime. After reaching the primary target composition, the oxidizing slag is removed from the furnace, and liquid steel is initially deoxidized. After that, the required steel composition is "built up" by adding alloying elements (mainly FeCr) into the furnace. The advantage of this technology is that it achieves a steel composition with low contents of impurities such as P, As, Sn, and Zn. The practice is used for the production of high-Cr tool steel grades (and stainless) with low P requirement. Oxygen injection after FeCr addition into the EAF may be required to reduce the contents of silicon and/or carbon before tapping. The disadvantage of "build-up heat practice" is extended furnace tap-to-tap time, which means reduced EAF productivity and higher costs for alloying elements instead of using scrap.

2.1.1.4.3 Recovery Heats

This practice is used to recover the easily oxidizing elements (mainly Cr, W, V, etc.) contained in the scrap. The melting process is conducted without oxidizing period under reducing slag and, in general, with a limited use of oxygen. The chemical composition of the steel during the meltdown period is controlled. In this case, it is adjusted by blending different scrap qualities of purchased and return scrap and, if necessary, alloys. Because of the reducing conditions, phosphorus removal from steel is not possible; therefore, the furnace charge mix must be composed of relatively clean scrap with known composition. The result of this process is an increased specific furnace capacity. Since foaming slag practice cannot be used, melting power input has to be set up with shorter arcs.

2.1.2 SECONDARY METALLURGY

Secondary metallurgy is a very important process step in "clean" steel production with respect to the nonmetallic inclusion level, gas content, and a close chemical analysis. The typical process route in the secondary metallurgical treatment of tool steels is the production via the ladle furnace (LF) and the vacuum degassing (VD) process followed by the casting process. There exists a possibility to produce tool steels via AOD (argon oxygen decarburization), LF, and then casting, but this route is not commonly used in tool steel production and only some special steel producers are following this (e.g., Böhler special steel). Therefore, this section gives a brief overview of the basic principles and most important considerations of the LF and VD furnaces.

Secondary metallurgy begins with the tapping of heat from the EAF. During tapping, deoxidation, first with carbon and afterward with aluminum, is done. Second, the alloying elements such as ferrochromium, vanadium, and other stable alloying elements are added to the heat. In addition, lime (CaO) is added in order to build up the metallurgical reactive slag. The typical slag consists of CaO, alumina (Al_2O_3), and silica (SiO_2). For protecting the ladle lining, a small amount of magnesia (MgO) is also added. The ternary phase diagram of the system $CaO–SiO_2–Al_2O_3$ indicating typical composition of the used slags and the nonmetallic inclusions is shown in Figure 2.5.

After the initial deoxidation and the slag creation, the heat is transferred to the ladle furnace. A sketch of a state-of-the-art LF can be seen in Figure 2.6.

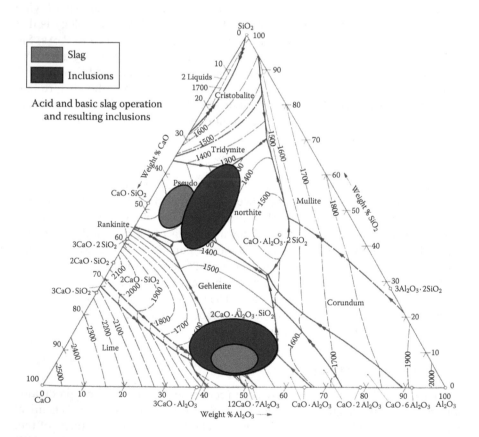

FIGURE 2.5 Ternary phase diagram of the $CaO–SiO_2–Al_2O_3$ system, indicating typical composition of the slag and the nonmetallic inclusions. (From Verlag Stahleisen mbH, 2nd edn., Ed. Verein Deutscher Eisenhuttenleute, 1995, 636pp.)

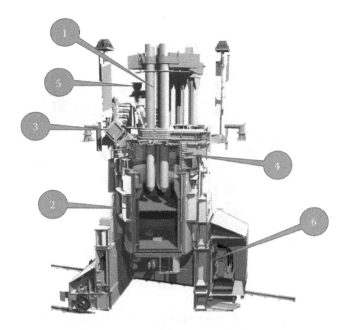

FIGURE 2.6 Sketch of a state-of-the-art ladle furnace with three electrodes for heating, argon stirring system, and alloy addition system. Codes: (1) Electrodes, (2) ladle, (3) operator gate, (4) roof, (5) alloy addition, and (6) transfer car.

The main tasks of this aggregate are the following:

- Deoxidation, that is, removing the residual oxygen content
- Homogenization by argon or inductive stirring
- Desulfurization
- Alloying, for adjusting the right chemical composition
- Heating

All these process steps are necessary to set the right chemical composition, cleanliness, and temperature before the vacuum treatment.

A precondition for proper deoxidation and desulfurization reaction in the LF is the adjustment of the slag. A well-deoxidized slag is necessary to produce clean steel. An indication for that is the color of the slag, which is visually checked whether the slag has a white color (so-called "white" slag) during the entire process. If the color is green, gray, or black, fine-grained aluminum should be added to deoxidize the slag. Therefore, precaution should be taken so that no or only limited air can enter the process, which means that, especially, the sealing between ladle and cover should be regularly checked.

To reduce the content of residual gases (nitrogen and hydrogen), a vacuum treatment is necessary. One particular case is certain tool steels for large forged blocks (e.g., plastic mold steels) in which the hydrogen content in the final product needs to be kept ≤2 ppm. The reason is that tool steels are very prone to hydrogen flacking.

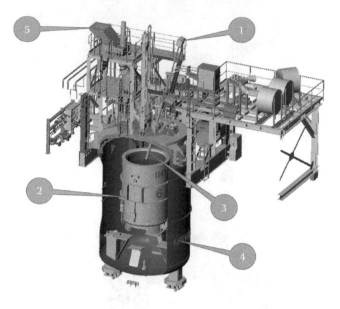

FIGURE 2.7 Sketch of a state-of-the-art VD plant with temperature and sampling lance and auxiliary equipment such as vacuum system and alloying system. Codes: (1) Temperature and sampling device, (2) ladle, (3) wire feeding, (4) vacuum vessel, and (5) alloy addition.

During the casting process, in general, ~0.5 ppm hydrogen pickup occurs. This means that the hydrogen content of the liquid metal after the vacuum treatment shall be ≤1 ppm in order to reach the required final hydrogen content in the finished product. In case of tool steels, the VD plant is commonly used. A scheme of a state-of-the-art VD plant is shown in Figure 2.7. By exposing the liquid metal to an atmosphere under low pressure, the dissolved gases are easily vaporized. During the VD process, stirring with argon also takes place to increase the speed of degassing reactions.

To modify the type of inclusions from Al_2O_3 (Type B) into a combined type of $CaO–Al_2O_3$ (Type D), the addition of calcium in the form of CaSi wire should be done as the last step of the secondary refining process. The final treatment step in the VD process is soft stirring with argon, which promotes the precipitation of nonmetallic inclusions in the slag. The heat is now ready for the subsequent casting process.

If, for any reasons, the temperature of the heat after the vacuum treatment is too low, a second LF treatment might be necessary in order to bring the heat up to the right casting temperature.

2.1.3 INGOT CASTING

After finishing the preparation of liquid steel by secondary metallurgical treatment (ladle furnace and vacuum treatment), the deoxidized and desulfurized "clean steel" is cast into defined shapes, dimensions, and weights.

Ingot casting is today especially used for the following:

- Smaller ingots (usually below 5 tons) for further hot deformation by rolling
- Larger ingots, with up to 100 tons, to be forged into bigger dimensions.

Due to the fact that most tool steels are produced by ingot casting, more details on this process are given in Sections 2.1.3.1 through 2.1.3.4.

2.1.3.1 Shapes of Ingots

In general, the liquid steel is cast in different shapes such as the following:

- Square or rectangular ingots for further rolling
- Octagonal or polygonal ingots, mainly for further forging
- Round ingots for tube production or as electrodes for electroslag remelting (ESR) and vacuum remelting (VAR)

The critical parameters for the mold design are the ratio of the height to the diameter (well known as the $H{:}D$ ratio) and the taper of the mold. Especially in the case of alloyed special steels (such as tool steels), the taper is the main factor that influences the internal soundness of the cast product. And in terms of segregation, the design and shape of the mold are crucial aspects to achieve a good-quality cast ingot. The ratio H/D ratio for forging ingots, depending on the size, will be in the range 0–2.0 and the taper in the range 9%–16%; for example, for a 12 ton H13 ingot, the $H{:}D$ is ≤1.4 and the taper 15%.

2.1.3.2 Types of Ingot Casting

There are two different setups to cast an ingot: top-pouring or bottom-pouring (see Figure 2.8a). In the top-pouring process, the molds are directly filled from the

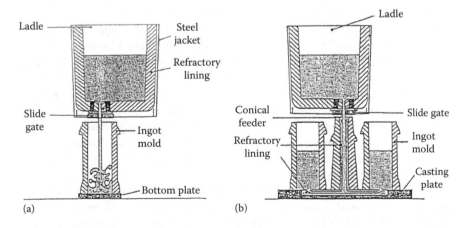

(a)   (b)

FIGURE 2.8   Sketch of (a) top-pouring and (b) bottom-pouring for ingot casting. Because of the more homogeneous filling of the mold and less splashing, bottom-pouring is the casting method of tool steels. (From Jellinghaus, M., *Stahlerzeugung im Lichtbogenofen*, 3rd edn., Stahl Eisen, Düsseldorf, Germany.)

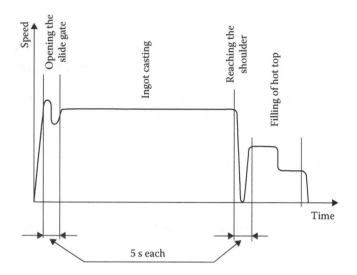

FIGURE 2.9   Casting speed variation versus casting time in a bottom-poured ingot.

top of the mold without any equipment and refractory between the ladle and the mold. Because of casting splashes, which adhere between the ingot and the mold wall, the surface of an ingot in top-pouring condition is in general not very smooth. Furthermore, prolonged contact with the oxygen in the air during splashing leads to an increase in reoxidation and inclusion levels. Therefore, the top pouring is not used for tool steels. In bottom pouring, on the other hand, the liquid steel is homogeneously filled through the bottom of the mold (see Figure 2.8b). The bottom-pouring method covers the range of 0.5 ton up to ~100 ton ingots, for rolling and forging. Bottom pouring allows not only casting single ingots, depending on the ingot size, but also producing up to 16 ingots per bottom plate (Figure 2.9).

To achieve proper quality, control of the reaction of the liquid steel with the refractory material and a constant speed are important factors for the internal and surface quality of ingot cast products. The refractory material used for this casting system should be of high quality, for example, ≥60% Al_2O_3, to prevent erosion by the relatively high casting speed of the liquid steel and therefore prevent exogenous inclusions. In terms of casting speed, in bottom-pouring of tool steels, the liquid steel must rise slowly and constantly in the mold and is always covered by a layer of casting powder to prevent contact with the atmosphere. An example is shown in Figure 2.8, where the speed is constant over the entire casting process and then reduced for filling the hot top part.

Auxiliary material such as casting powder or additional equipment such as hot top insets can be used in ingot casting, to improve the quality of the as-cast ingot. Casting powder is added to the top surface of the liquid metal. The powder melts as a result of the high temperature of the liquid steel and penetrates into the gap between the ingot mold wall and the cast ingot, forming a vitreous layer and generating a smooth as-cast surface. The selection of a proper casting powder depends on the grade of the cast steel. They differ mainly with regard to the different carbon content, from 0.5% up to 15%C, to regulate the melting point and guarantee the required surface quality of the as-cast ingot. This is a crucial factor for the subsequent hot deformation to avoid the need for ingot surface conditioning by grinding, thereby increasing the overall yield.

The second important practice concerns the use of hot top inserts. During solidification, the ingot body shrinks and the liquid may not be able to feed all regions. Voids may then be formed in the regions that solidify last, such as the top of ingots. In order to minimize those voids, hot tops and hot top treatment may be used. The hot top insets may be either inserted in the mold or placed on top of the mold itself, as shown in Figure 2.10a and b, respectively. In both cases, insulating materials are used to reduce the heat transfer through the mold through the cast iron ring of the hot top, which is placed on the mold, as can be seen in Figure 2.10c.

In addition to the insulating plates for radial insulation on the surface, exothermic and insulating powders are often used to keep the hot top area liquid as long as possible, so that the top of the ingot can be filled with liquid steel to reduce the shrinkage cavities. After filling the hot tops, while the steel is still covered with casting powder, the exothermic powder is immediately added, and the heat of reaction generated by the exothermic powder with the steel heat of burning, for example, fuse of metallic aluminum, heats up the remaining liquid steel again. Furthermore, insulating powder is added also to prevent extensive heat loss.

2.1.3.3 Advanced Teeming Systems (ATSs)

A fairly new process for ingot casting is the ATS system (advanced or semiautomatic teeming system) [3]. This teeming system is based on a semiautomated circuit of the involved casting equipment, as is schematically illustrated in Figure 2.11. It consists of several stations where the teeming sets are put together and prepared for casting. The main characteristic of this system is that the preparation area and the stripping are separated from the casting area of the ingots. Some process steps in ingot casting normally done manually or carried out by machines. These process steps are cleaning the bottom plates, implementing the runner bricks, and cleaning the molds. This system is preferred for smaller ingot sizes in the range of 1–6 tons so that a large number of molds can be used.

2.1.3.4 Solidification

After the casting process, the liquid steel begins to solidify. During this process, absolutely no movement of the mold should take place. The solidification time can be calculated depending on the largest diameter according to the following equation:

$$s = c * \sqrt{t}$$

where
   s is the diameter (in mm)
   c is the solidification constant
   t is the solidification time in minutes

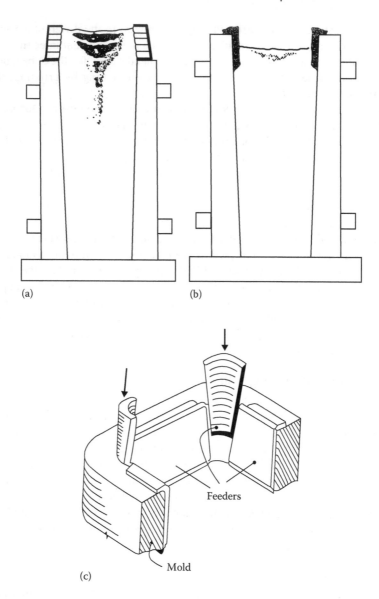

FIGURE 2.10 Different types of hot tops: (a) on top of the mold, (b) inserted in the mold, (c) feeders placed in the mold.

After the ingot is totally solidified, it can be stripped. Especially round ingots have to be stripped immediately after complete solidification. Depending on the next planned production step, the ingots are transferred in hot condition to the forging or rolling mill, or cooled down to room temperature. The hot transfer is done either with the ingot still in the molds and stripped on site, or stripped in the casting pit and transferred in insulating boxes, brought then to the forging or rolling mill, and

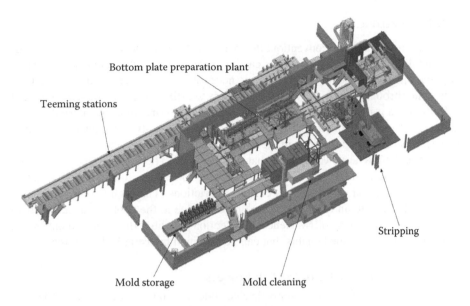

Bottom plate preparation plant

Teeming stations

Stripping

Mold storage   Mold cleaning

FIGURE 2.11   State-of-the-art semiautomated ingot teeming system.

immediately charged in furnaces at ~600°C. The cool-down period depends mainly on the carbon content, and cooling can be done by either cooling down to room temperature, slowly cooled in insulated boxes, or annealed in furnaces and then cooled down.

2.1.4 CONTINUOUS CASTING

The continuous casting process is the most popular in carbon steels and low-alloy steels. High-alloyed tool steel grades are difficult to cast in standard bow-type continuous-casting machines. This is because the risk of formation of surface and internal defects during strand bending, unbending, and straightening is very high. In addition, the high carbon and high alloy content of tool steels may lead to strong segregation after continuous casting, thus limiting the application of this method in several products. Nevertheless, alternative continuous casting methods were developed and are currently applied by various steel producers. These methods are horizontal continuous (HCC), vertical continuous (VCC), and segment (SC) casting.

The combined use of the moveable electromagnetic stirrer and vertical continuous casting machine allows getting a larger equiaxial zone and low segregation content compared to other production methods. The use of each of the production methods depends mainly on the customer requirements, production mix, and quantity, as well as the investment in equipment.

However, even with newly developed technologies that enable a better control of casting parameters and segregation, continuous casting is still not common in tool steel production. So ingot castings still represent the main route for tool steels, especially for larger diameters.

2.1.5   REMELTING

Although nowadays the conventionally cast tool steels produced via EAF, secondary metallurgy, and ingot continuous casting have a high product quality, for certain applications an additional process step is necessary to enhance the quality further. It is done through remelting, where an already solidified ingot is melted again and solidified in a water-cooled copper mold, enhancing in particular the internal quality and as-cast structure of steel (less segregation, good density, etc.), thus resulting in superior mechanical properties in the final product.

ESR and VAR are two options that can be used. The application of the VAR process is quite limited for tool steels (except for some special applications), as this process for high-carbon steels has certain restrictions. The high carbon content in some steels leads to an intensive boiling reaction due to the applied vacuum, which in turn affects the arc stability. Hence, this section describes the basic principles and gives an overview of the benefits that can be obtained by using ESR refining.

2.1.5.1   Basic Principles of the ESR Process

The ESR process utilizes a consumable electrode (which is an ingot or a continuous-cast ingot), which is melted through a superheated metallurgical reactive slag layer into a water-cooled copper mold applying alternating or—for larger sections—low-frequency current (0.2–0.4 Hz). The function of the slag is not only to generate heat but also to transfer the heat energy to the electrode (for melting), to the liquid metal leaving the electrode (for superheating), and to the ingot surface for maintaining the required temperature gradients in the solidifying ingot [4,5]. As the slag temperature rises above the melting point of the metal, the tip of the electrode submerged in the slag bath melts. The molten metal droplets fall through the liquid slag and are collected and solidified within the liquid metal pool. A schematic representation of the ESR process is given in Figure 2.12.

The intensive reactions between the metal droplets and the slag result in a refining process of the steel. The entire ESR process is contained inside a water-cooled copper mold, which removes the heat from the ingot, causing the metal in the liquid pool to solidify. The cooling also causes a small portion of the slag to solidify against the mold wall, forming a thermal and electrical insulator. This barrier is referred to as the *slag skin*. The slag skin thickness, which is affected primarily by the slag chemistry, melting rate, and immersion depth of the electrode, has a large impact on the surface quality of the remelted ingot [5]. The surface quality of an as-cast ESR ingot is, in general, smooth, and allows a direct hot working operation without any surface conditioning.

Nowadays, all state-of-the-art ESR furnaces are equipped with a protective gas hood, which ensures that no ambient air can enter the process. Usually argon, nitrogen, or a mixture of both gases is used as the protective gas. This ensures a well-controlled furnace atmosphere and prevents undesirable oxidation of the electrode surface that can lead to poor cleanliness levels. One of the main objectives is to produce ingots with a sound, homogeneous structure that is free of macroscopic defects such as inclusion, macro-segregations, or voids. By utilizing optimum remelting parameters and observing good process control, the ESR process

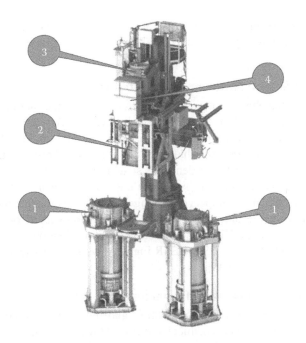

FIGURE 2.12 Schematic view of the ESR process. Codes: (1) Molds, (2) protective gas hood, (3) furnace head, and (4) slag and alloy addition.

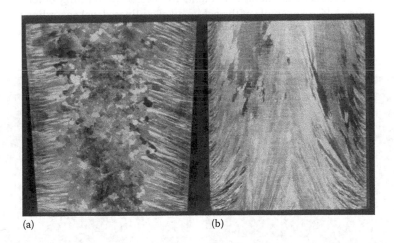

(a)   (b)

FIGURE 2.13 As-cast structure of (a) a conventionally cast ingot and (b) an ESR ingot. (From Krainer, E. et al., *BHM*, 116(3), 78, 1971.)

can produce ingots with a directionally solidified primary structure over the entire ingot cross section [6].

Figure 2.13 shows a comparison of the typical solidification structure of a conventionally cast ingot and an ESR ingot [7]. As can be seen in the images, the ESR ingot has a directional solidification structure from bottom to top and from the surface down to the middle, whereas in the conventional ingot, due to the change of the heat balance during solidification, also a globular equiaxial structure in the middle is evident.

2.1.5.2   Solidification Structure of Remelted Tool Steels

The structure of remelted ingots is mainly determined by the melting rate, cooling rate, slag temperature, and the direction of solidification. Basically, it can be stated that a shallower liquid metal pool results in better internal quality of the remelted ingot. The lower the melt rate, the shallower the liquid metal pool depth. Figure 2.14 shows a comparison of different dendritic structures of a 550 mm diameter ESR ingot of H11 [8]. As is evident, the higher the melt rate, the coarser the solidification structure, which has a direct impact on the ingot quality. Another example is given in Figure 2.14, showing the annealed microstructure of a hot-worked tool steel that was analyzed according to the VDEh rating chart for the evaluation of the macrostructure. As is evident in Figure 2.15, the ESR ingot shows a finer structure compared to the conventionally cast ingot [9].

2.1.5.3   Metallurgical Quality of Remelted Tool Steels

When tool steels are remelted, solidification and ingot structure control are considered more important than chemical refining. However, a considerable refining potential is also given. ESR slags contain fluorspar, lime, and aluminum oxide, as well as small amounts of magnesium oxide, silicon oxides, and other components. In general, the higher the fluorspar content, the better the refining potential of the slag. In Figure 2.16, an example of the achieved cleanliness level according to ASTM E45 Method D of H11 (X38 CrMoV51) is given after different process routes [10].

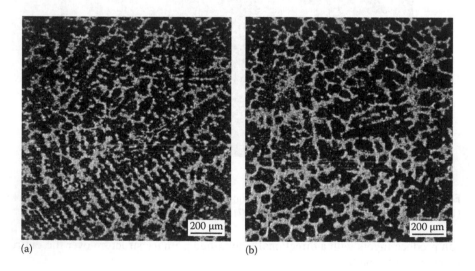

(a)   (b)

FIGURE 2.14   Dendritic structure of a 550 mm diameter H11 (X38CrMoV 51) ESR ingot at different remelting rates: (a) 390 kg/h and (b) 540 kg/h. (From Holzgruber, W., *Radex-Rundschau*, 3, 409, 1975.)

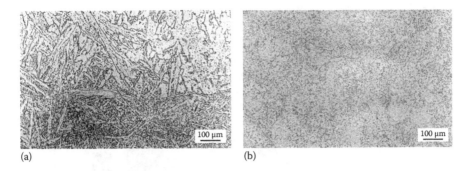

FIGURE 2.15 Microstructure evolution in the annealed condition: (a) conventionally cast ingot, 100×, according to SEP1614; GB 3 and (b) ESR remelted ingot, 100×, according to SEP1614; GA1. (From Ghidini, A. and Boh, M., The new Lucchini Siderneccanica ESR plant and the manufacturing of special AISI H11 and H13 hot work tool steel grades, *Inteco Symposium*, Chicago, IL, 2002.)

2.1.5.4 General Remelting Practice for Tool Steels

Taking all the earlier-mentioned considerations into account, there are certain restrictions in ingot size and remelting parameters that should be kept in mind. As the ingot structure is one of the most important parameters for remelting tool steels, the remelting rate is the predominant factor to achieve dense, sound, and segregation-free, ingot structure. As a rule of thumb, the melt rate for tool steels, in kg/h, may be calculated multiplying the diameter in mm by 0.8–1.0; for example, when producing a 1000 mm diameter ESR ingot, the remelting rate should be in the range of 800–1000 kg/h. The right choice of the melt rate mainly depends on the segregation behavior of the remelted tool steel and the used mold size. Typical slags contain a fluorspar content of 30%–50% depending on the desired refining potential.

A very important precondition is the quality and hardness of the electrode. As there exists a very steep temperature gradient over the entire length of the electrode, the possibility of the electrode cracking during remelting is high if the electrode is too hard. The electrode hardness should be in the range of 260–280 HB, which means that a soft annealing before remelting takes place might be necessary.

The choice of the used protective gas mainly depends on the nitrogen content of the remelted grade. If no nitrogen pickup during remelting takes place, 100% argon is used. In case only argon is used, special care should be taken for the immersion depth of the electrode. Too shallow an immersed electrode can lead to arcing, which disturbs the process control.

Controlling the process is one of the most important factors influencing the ingot quality. The melt rate should be stable during the entire remelting operation, as very high fluctuations of the melt rate can lead to internal defects in the ingots.

Depending on the grade, the finished remelted ingot is cooled down in air, in a cooling box, or in the furnace, or is brought directly to the heating furnace and brought to the forging temperature.

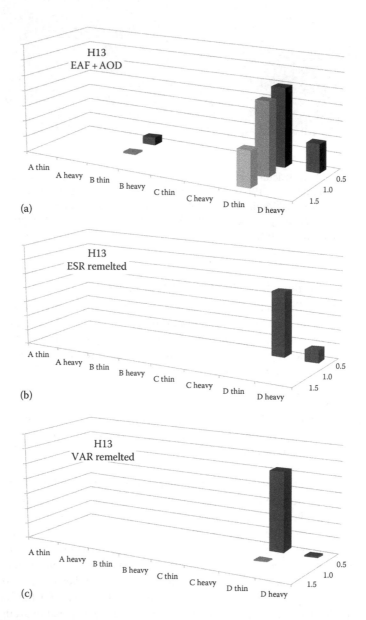

FIGURE 2.16 Achievable cleanliness level according to different production routes, for H13 tool steels: (a) H13 steel produced by EAF + AOD, followed by conventional casting and forging. (b) Same as (a), but with remelting by ESR after casting and prior to forging. (c) Same as (a), but with VAR remelting after casting and prior to forging. (From Reiter, G. et al., The influence of different melting and remelting routes on the cleanliness of high alloyed steels, *LMPC*, 2013.)

2.1.6 Powder Metallurgy

Powder metallurgy has become a major process in the manufacture of improved cold-work and high-speed tool steels in the form of as-compacted or hot-worked billet and bars, semifinished parts, near-net shapes, and indexable cutting tool inserts. Because of the refined microstructure, when compared to conventionally produced high-alloy grades, tool steels produced via this production route have significantly increased wear resistance, toughness, ductility, and fatigue life. Further, the alloying flexibility of powder metallurgical tool steels allows the production of new tool steels that cannot be produced by conventional ingot processes due to segregation-related problems. Rapid solidification of the produced powders used for powder metallurgical tool steels eliminates such segregation and produces a very fine and homogeneous distribution of carbides.

The basic principle in the manufacture of tool steels and HSS is preparing a prealloyed melt and turning the molten metal into powder through a gas atomization process. The resulting spherical powder is collected and encapsulated in metal containers, heated to forging temperatures, and transformed into fully dense billets via hot isostatic pressing (HIP). In most cases, the billets are then rolled or forged to various bar sizes [11]. Figure 2.17 shows a schematic illustration of the material flow during the production of high-performance tool steels.

Although the first generation of powder metallurgical tool steels already showed improvements over conventionally produced tool steels, a variation in performance mainly due to the rather high nonmetallic inclusion content was noticeable. In powder metallurgical tool steels, carbide size is usually 2–4 µm while the nonmetallic inclusion size can be 5–15 µm, making the nonmetallic inclusions the primary defect in the microstructure [12]. As the steel cleanness has been identified as a critical factor in tool performance, manufacturers of powder metallurgical tool steels have focused on reducing the nonmetallic inclusion content in the alloys.

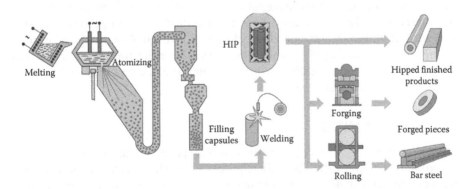

FIGURE 2.17 Production route for high-performance tool steel. (Adapted from Tornberg, C. and Fölzer, A., New optimised manufacturing route for PM tool steels and high speed steels, In *Proceedings of the Sixth International Tooling Conference—The Use of Tool Steels: Experience and Research*, Eds. J. Bergstrom, G. Fredriksson, M. Johansson, O. Kotik, and F. Thuvander, Karlstad, Sweden, pp. 305–316, 2002.)

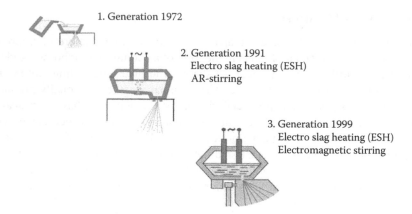

1. Generation 1972

2. Generation 1991
Electro slag heating (ESH)
AR-stirring

3. Generation 1999
Electro slag heating (ESH)
Electromagnetic stirring

FIGURE 2.18 Development of the manufacturing process for tool steels. (From Böhler Edelstahl GmbH & Co KG, http://www.bohler-international.com/english/files/download/ ST035DE_Microclean.pdf.)

Since the establishment of the powder metallurgical production route for tool steels in the 1970s, numerous enhancements have been developed over the years to keep up with the demands placed on the products during application. A schematic illustration of the development of the manufacturing process for powder metallurgical tool steels is given in Figure 2.18. The traditional production process employed a ladle and a small tundish, which was refilled constantly during atomization. Because of the high risk of slag entrainment during refilling and the resulting high amount of nonmetallic inclusions in the produced powder, the tundish size was increased to require less refilling. The tundish was heated via electroslag heating (ESH), and inert gas stirring was applied to keep the alloying elements evenly distributed during the atomization process. Even though the steel cleanness was increased compared to the that obtained by the traditional method, the inert gas passing through the slag layer still introduced slag (nonmetallic inclusions) into the molten metal. Today, inert gas stirring is replaced by electromagnetic stirring for melt homogenization without disturbing the slag layer, thereby further reducing the amount of nonmetallic inclusions in the produced steel powder.

2.2 HOT FORMING

After the several possible melt preparation and cast processes, the ingots are ready to be formed in the solid state. The main requirements in plastic deformation processes are the elimination of macro- and microporosities, the refinement of the microstructure, and, thereby, the improvement of the mechanical properties as well as the shaping of the material to its final dimension. Usual deformation processes are forging and rolling, though in some cases extrusion or drawing is also applied. Most of the deformation processes are hot-forming processes, but especially for thin sections such as strip steel or wires, also cold-forming processes can be used.

2.2.1 Metallurgical Aspects

Besides the shaping of the material, plastic deformation of various grades of tool steels also provides complex metallurgical relationships to be considered. We give just two examples.

In ledeburitic steels such as high-carbon, high-chromium cold-work tool steels (e.g., D2, D6/X153CrMoV12-1, X210CrW12) or high-speed steels (e.g., M2/HS6-5-2), it is impossible to dissolve all carbides in the solid state. Therefore, the initial carbide network from the solidification process can be broken and refined only during the forming process. Plastic deformation therefore always takes place in a two- (or more) phase region, which includes at least one phase (the carbides) that cannot be deformed in practice. This requires special care for the selection of temperature and deformation rates during the single deformation steps. Nevertheless, even at significant degrees of deformation, the differences in carbide size over the cross section cannot be completely eliminated. Figure 2.19 shows an example of the carbide distribution in a 60 mm bar of the steel D6/X210CrW12 in the longitudinal direction (left: center; right: near the surface) with significantly coarser and more bent carbide structures in the center regions.

In matrix-type tool steels such as hot-work tool steels (e.g., H11/X38CrMoV5-1) or (corrosion-resistant) plastic mold steels (e.g., 420/X42Cr13), a very fine microstructure is requested over the whole cross section. Such microstructure is usually achieved by a tight control of carbide precipitation and deformation toward the end of the forming process. Because of the limitation in thermal conductivity, this is a significant challenge for the process control, especially at larger cross sections that are typical for these types of steels.

The effect of deformation is also evident in the material properties after the final heat treatment. Figure 2.20 shows the tensile strength, elongation to fracture, and impact toughness (DVM-specimen) for a conventionally produced (ingot-cast) and ESR hot-work tool steel (~H11/X40CrMoV5-1) in the center region for different degrees of deformation. Besides the beneficial effect of the ESR process, it can be seen clearly that a minimum degree of deformation of ~4 is necessary to achieve

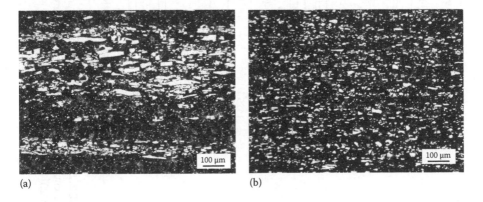

(a)   (b)

FIGURE 2.19 Carbide morphology in a soft-annealed, 60 mm steel bar of the steel D6/X210CrW12: (a) center region and (b) near the surface.

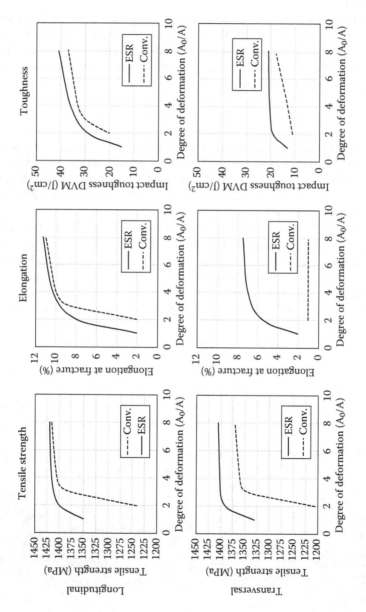

FIGURE 2.20 Material properties (tensile strength, elongation to fracture, and impact toughness) of the steel ~H11/X40CrMoV5-1 in the center region for different production routes and various degrees of deformation (ratio of cross section: A_0/A). (Adapted from Randak, A. et al., *Stahl Eisen*, 92(20), 981, 1972.)

properties typical for the capabilities of the material. Nevertheless, elongation and toughness in the transverse direction do not reach the longitudinal values even at much higher degrees of deformation.

2.2.2 HOT FORGING

While forging rather than rolling is usually applied to larger sections, a wide area of overlap exists where both technologies are applied. For very heavy sections, but also for bar material with larger cross sections, forging presses offer high flexibility. In contrast, radial forging machines provide high productivity for bar material with medium to high cross sections. Usually, forging presses have the advantage that plastic deformation extends over the whole cross section even if lower degrees of deformation are applied.

For very large cross sections, sometimes in the range of 1000 mm, a sufficiently high degree of deformation cannot be reached by a single direction of deformation. Therefore, press-stretch forging or three-dimensional forging is applied. For the planning and control of such forging operations, numerical simulation methods have found widespread application.

2.2.3 HOT ROLLING

Rolling operations are usually applied for the production of long and flat products, starting from medium sections down to wires, bars, and plates with less than 10 mm thickness/diameter. Therefore, different kinds of rolling plants (e.g., cogging mills, multiline rolling mills, plate mills) exist. Usually, simple cross sections (round, square, rectangular) are produced, but also more complex near-net-shape profiles for certain applications are possible.

2.3 HEAT TREATMENT

Heat treatment during production can be performed in separated processes, but in consideration of the general requirement to reduce energy consumption, it is often integrated in forming operations. Diffusion annealing can be placed ahead of the (first) forging or rolling step, thereby replacing the (anyway) necessary heating to deformation temperature. In contrast, soft annealing can be performed at the end of the forming process as a controlled cooling operation. More details on the heat treatments performed by the steel producers are given in Section 3.2.1 of Chapter 3.

2.3.1 DIFFUSION ANNEALING/HOMOGENIZING

The solidification of tool steels always leads to a segregation of the alloying elements. The larger the solidification interval, the more pronounced such segregations. This parameter is determined by the alloy composition and cannot be avoided. The second important aspect in solidification is the solidification time, which also contributes to the severity of segregations. While there is almost no segregation in powder-metallurgical products, ingots with large cross sections are prone to significant segregation effects. Therefore, two different types of segregation need to be differentiated: first, ingot segregation (in some references also named macrosegregation), and, second, crystal segregation (also named microsegregation).

Ingot segregations are the result of fluid flow and shrinkage effects during solidification and spread over the whole length and cross section of the ingot. This type of segregations can be overcome only by remelting processes such as ESR and VAR, as the distances are much too wide for diffusion-controlled heat-treatment processes.

Crystal segregations are the result of the dendritic growth of the solidification front. The distance (dendrite arm spacing) between enriched and depleted areas within a crystal is in the range of 0.01 mm up to about 1 mm and depends strongly on the local solidification rate. Therefore, center regions of large ingots have the largest dendrite spacing, and it is most critical to eliminate this type of segregation. Remelting process can significantly reduce the dendrite spacing by about one order of magnitude but cannot eliminate them completely [15].

Not all chemical elements show the same degree of segregation. While the usual degree of segregation (maximum/minimum concentration) for chromium is in the range of 2, molybdenum and phosphorus can reach a degree of segregation up to 10 [15,16]. Also, carbon shows a high degree of segregation, but due to the much higher diffusion rate of carbon compared to other alloying elements, homogenization is usually achieved under normal production conditions without special heat treatment.

Segregations, as local inhomogeneity of the chemical composition, lead to the formation of inhomogeneous microstructures (e.g., carbide stringers, different martensite/bainite ratios) and thereby to inhomogeneous and anisotropic material properties. From the thermodynamic point of view, they are not in equilibrium, and therefore tend to be reduced, provided adequate activation energy (temperature) and time are given. Any application of high temperatures to the material, such as heating for hot-forming, contributes to a reduction of segregations, but for a better reduction, longer holding periods at high temperatures are necessary. The maximum temperature for this heat treatment is limited either by the melting point of the lowest melting (segregated) phase or the capabilities of the furnace, and is usually in the range of 1100°C–1300°C (2010°F–2370°F). The holding period is economically limited, as the rate of improvement shrinks with longer treatment times. Usually, holding periods of about 1 day (24 h) are applied.

2.3.2 SOFT ANNEALING/SPHEROIDIZING

Soft annealing is usually performed at the end of the production process of tool steels to provide the best possible microstructure to the machining of the tools, eventually also to improve the cold formability. The main metallurgical aspect of soft annealing is, therefore, the formation of spheroidized carbides. This can be achieved either by a long holding period just below the austenitizing temperature A_{c1} (Figure 2.21a), or by a slow and controlled cooling process starting just above A_{c1} (Figure 2.21b).

The first type of heat treatment (Figure 2.21a) is typical for low-C high-Ni containing tool steels (e.g., 6F5, 6F7/45NiCrMo16, 50NiCr13) or for low-alloyed tool steels in the normalized starting condition. High nickel alloyed steels transform to martensite even when slow air cooling or furnace cooling is applied. Therefore, a soft microstructure can be achieved only by tempering at high temperatures just

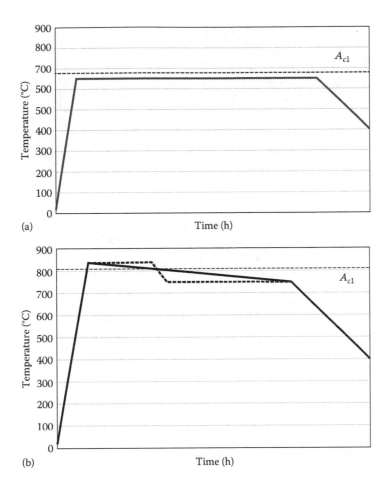

FIGURE 2.21 Soft annealing of tool steels: (a) without prior austenitization and (b) with prior austenitization.

below the formation of fresh austenite (A_{c1}). The microstructure consists of the ferrite matrix (tempered martensite) with small, uniformly distributed temper carbides (Figure 2.22a).

The second type of heat treatment (Figure 2.21b) is more general and used for all high(er) carbon containing grades. Therefore, the material is first heated above A_{c1} to form austenite. Secondary or ledeburitic carbides should not be dissolved. These carbides (especially the secondary carbides) get spheroidized during the first holding step. During the subsequent slow cooling (or isothermal holding after a short quenching to just below A_{c1}), the austenite transforms in a pearlitic reaction to a microstructure with small globular carbides. Figure 2.22b shows such a microstructure of fine globular carbides and some elongated gain boundary (i.e., secondary) carbides as examples of the soft-annealed steel S1/60WCrV7. In the last picture, a soft-annealed D6/X210CrMoV12 (Figure 2.22c) with additional large ledeburitic carbides can be seen.

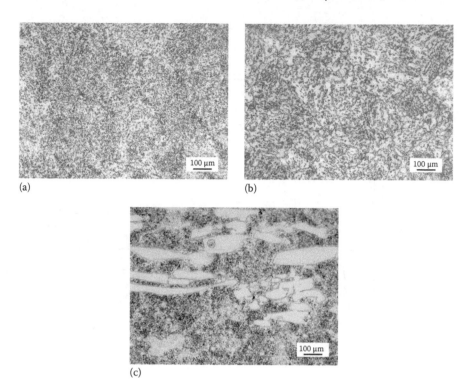

FIGURE 2.22 Soft-annealed microstructures of different tool steels: (a) 50NiCr13, (b) S1/60WCrV7, and (c) D6/X210CrMoV12.

ACKNOWLEDGMENT

The authors gratefully acknowledge the support of INTECO melting and casting technologies GmbH throughout the process of writing this chapter. Special thanks to professionals who contributed their time and expertise and for giving their feedback and comments to the authors of this chapter.

REFERENCES

1. Verlag Stahleisen mbH, 2nd edn. Ed.: Verein Deutscher Eisenhuttenleute, 1995, 636p.
2. M. Jellinghaus. *Stahlerzeugung im Lichtbogenofen*, 3rd edn. Düsseldorf, Germany: Stahl Eisen, 1994.
3. C. Redl, S. Hammarqvist. Design and application of advanced semi-automated ingot teeming systems. *Proceedings of AIStech*, Pittsburgh PA, 2013.
4. A. Mitchell. Slag functions in the ESR process. *Proceedings of the International Symposium on Liquid Metal Processing and Casting*, Santa Fe, NM, 2005.
5. G. Hoyle. *Electroslag Processes—Principles and Practice*. London, U.K.: Applied Science Publishers Ltd., 1983.
6. H. Holzgruber, W. Holzgruber. New ESR concepts for improved control of solidification. *Proceedings of the 2002 International Symposium on Electroslag Remelting Technologies*, Chicago, IL, 2002, pp. 1–15.

7. E. Krainer, W. Holzgruber, R. Plessing. Praktisch isotrope Werkzeugstähle und Schmiedestücke höchster Güte. *BHM*, 116(3), 78–83, 1971.
8. W. Holzgruber. Möglichkeiten und Grenzen der Beeinflussung des Erstarrungsgefüges legierter Stähle beim Elektroschlacke-Umschmelzen. *Radex-Rundschau*, 3, 409–421, 1975.
9. A. Ghidini, M. Boh. The new Lucchini Siderneccanica ESR plant and the manufacturing of special AISI H11 and H13 hot work tool steel grades. *Inteco Symposium*, Chicago, IL, 2002.
10. G. Reiter et al. The influence of different melting and remelting routes on the cleanliness of high alloyed steels. *Liquid Metal Processing & Casting Conference 2013 (LMPC)*, 2013, Austin, TX.
11. C. Tornberg and A. Fölzer, New optimised manufacturing route for PM tool steels and high speed steels. *Proceedings of the Sixth International Tooling Conference— The Use of Tool Steels: Experience and Research*, Eds. J. Bergstrom, G. Fredriksson, M. Johansson, O. Kotik, F. Thuvander, Karlstad, Sweden, 2002, pp. 305–316.
12. B. Krause. What you don't know about P/M tool steel production that can affect productivity. *Stamping Journal*, 24–26, August 2004.
13. Böhler Edelstahl GmbH & Co KG. Available online at: http://www.bohler-international.com/english/files/download/ST035DE_Microclean.pdf, accessed March, 2016.
14. A. Randak, A. Stanz, W. Verderber. Eigenschaften von nach Sonderschmelzverfahren hergestellten Werkzeug- und Wälzlagerstählen. *Stahl und Eisen*, 92(20), 981–993, 1972.
15. H.-J. Eckstein. Wärmebehandlung von Stahl. Leipzig, Germany: VEB Deutscher Verlag für Grundstoffindustrie, 1971.
16. A. Schrader, De Ferri Metallographia. *Part 3: Erstarrung und Verformung der Stähle*. Eds.: A. Pokorny and J. Pokorny, Düsseldorf, Germany: Verlag Stahleisen m. b. H., 1967.

3 Physical Metallurgy of Tool Steels

This chapter is the core of the book, and was the first to be written. It represents the basis for tool steel behavior, focusing on the understanding of the properties and performance based on the microstructure. As detailed here, this reduces the complexity of the different types and distinct properties of tool steels. The basic concepts of ferrous alloys are not explained here (e.g., terminology, different phases, and other concepts of ferrous alloy metallurgy), but they can be found in basic materials science books, such as Reference 1.

3.1 REDUCING THE COMPLEXITY AND THE "MICROSTRUCTURE APPROACH"

The first time one starts studying tool steels, an immediate (and correct) conclusion will be, "tool steels are a complex type of materials," for many reasons. First, considering only the chemical composition according to ASTM of most common tool steels, the number easily reaches about 50 different grades (Table 1.1). In addition to their standard compositions, several companies develop their modifications, either with proprietary compositions or with modifications to the standard compositions. So, for simplicity, about 100 different grades are assumed to be present.

A second important characteristic of tool steels is their manufacturing process. For example, when evaluating casting methods of an H13 tool steel, it can be produced by electro-stag remelting (ESR) (see Chapter 2) when used in high-performance applications such as die-casting or plastic injection molds, or by casting using the conventional method, which is the common manufacturing condition for hot-forging dies. In addition to melt-shop processes, other forming or finishing conditions may change substantially depending on the application requirements of a given tool steel. Therefore, taking into consideration the total number of variables in the manufacturing conditions of the tool steel, in the steel mill the number can be very high, such as the following:

- Five melting technologies, with only ingot casting, considering two types of melting (electric arc furnace [EAF] and vacuum induction melting [VIM]) times two types of remelting (ESR or vacuum remelting [VAR]), plus the possibility of powder metallurgy.
- Two possibilities of hot-forming operation (forging or rolling).
- At least two different types of annealing treatment, leading to distinct microstructures or hardness.

- Three finishing conditions: machined (e.g., milled, or turned, ground) or rough/black surface.
- And in terms of delivery, the pieces can be cut from blocks or supplied from rolled/forged bars close to the final tool dimensions (two possibilities).

Therefore, in terms of manufacturing conditions, multiplying only the variables listed above, there are nearly 120 possibilities.

A third and very important characteristic of a tool steel is the final heat treatment, usually done by hardening and tempering. As shown throughout this chapter, the temperatures and times are chosen depending on the steel type and the requirements for applications, either by comparing the tempering charts found in reference books or (most common way) supplied by the tool steel manufacturer. There are several heat-treatment conditions, but for simplicity, let us consider only two hardening temperatures, five tempering temperatures, and five different heat-treatment times; this last parameter basically depends on the size of the tools and the heating furnace for a given tool steel. In addition, the cooling from quenching dramatically changes the quality of the final tool, since highly accelerated cooling may lead to distortion or cracks and slow cooling usually results in poor properties. Like the treatment time, the cooling condition depends basically on the furnace, the type of steel, and the dimensions/geometry of the tool. Again, for simplicity, we assume about five different cooling conditions here. So, combining all the variables (multiplication), there are about 250 distinct possibilities for heat treatment of a given tool steel.

To sum up, considering only the three main characteristics of a tool steel's definition (chemical composition, manufacturing method, and heat treatment), the number of possible variables is 100 (compositions) × 120 (manufacturing) × 250 (heat treatment) = 3 million! This is the degree of complexity to understand a given property or performance of a tool steel just using a direct association of variables, or, in other words, trying to understand their characteristics in terms of composition, manufacturing, and heat treatment. This, thus, explains why, at first sight, tool steels look (and are!) a complex group of metallic alloys.

The number of different combination also explains why it is difficult to establish direct correlation, especially in industrial trials, which is also intrinsically subjected to statistical variability. For instance, we may see situations where "steel composition A is changed to B and the tool life was double"; while the statement may be true, it is important to understand what the actual changes are, to elucidate whether the effect is due to the composition itself or to any other parameters. Very likely, steel A and B have different manufacturing methods (ingot size, forming temperatures, annealing cycle, and, possibly, finishing operations) or different heat treatments. So the improvement in performance cannot be solely attributed to the chemical composition, but it may be a result of all the differences that steel B presents in relation to steel A.

Therefore, the conclusion from this discussion is that there is a clear need to simplify tool steels to understand clearly the correlation between steel grades, heat treatments and manufacturing conditions, the final properties, and, as a consequence, the final performance. The theme of this chapter and of this book as a whole is to consider the *microstructure approach*. Especially for non-metallurgists

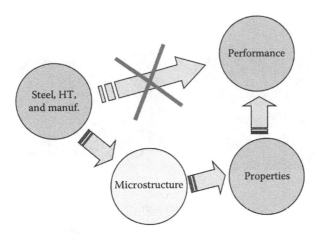

FIGURE 3.1 Microstructure approach, proposed in the present book. The representation shows the main tool steel variables, steel types, heat treatment (HT), or manufacturing, and that the direct correlation of those variables to the performance may be complex and misleading. The suggestion is that the variables are correlated to the microstructure, which may then explain the properties and finally the performance.

or nonmaterials engineers, the term "microstructure" immediately sounds complex and untouchable. It is usually easy to understand temperatures, treatment times, chemical composition, or even final properties, because all of them are given in numbers. We see a change from 1000°C to 1100°C, and the natural reaction is to try to correlate this 100°C difference to the final properties or performance. However, by doing this we are not dealing with one variable but with possibilities that can reach millions of combinations, as explained earlier. And a tremendous simplification can be achieved if the microstructure is considered. This change of rationale is shown in Figure 3.1. And understanding changes in microstructure is not as complex as it seems, as detailed in the upcoming sections.

As exemplified in Figure 3.2, the microstructure of all tool steels (all the three million possibilities) can be divided into just two general parts: the microstructural matrix and the undissolved (nonmetallic) particles. The microstructural matrix is the metal itself, in which most of the atoms of Fe and alloying elements are distributed. Actually, the microstructure of some tool steels (e.g., plastic mold steels and hot work steels) is designed based only on the microstructural matrix, although there are always present particles remaining from the manufacturing process, such as nonmetallic inclusions. However, the metallic matrix is not sufficient to respond in a proper way to all the requirements of a tool. The main example is the property of wear resistance, which requires constituents in the tool steel that are much harder than the metallic matrix (even after hardening to very high hardness). Thus, many tool steels are designed in a way that during solidification carbon and alloying elements combine to form nonmetallic particles (called *carbides*), whose hardness can be more than 10 times that of the matrix. The quantity, size, and distribution of those particles dramatically change the properties, usually improving the wear resistance and reducing the toughness.

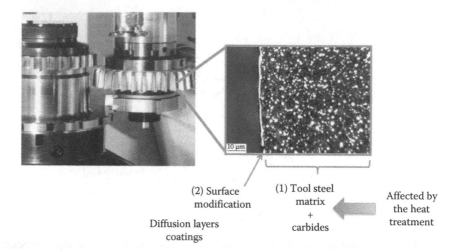

FIGURE 3.2 Typical microstructure of a high-speed tool steel, showing the proposed approach of matrix plus carbides.

Therefore, the approach to designing a microstructure of a tool steel can be compared to asphalt, where the solid thermoplastic mass (bitumen) is added to the mixture of stones and used as road paving material. The similarity is that asphalt may contain different amount of stones, which will then promote higher or lower resistance to wear. But, in asphalt, the particles are added to the bitumen to form the final mixture, whereas in tool steels the particles are formed naturally during solidification (in situ formation) from a homogeneous (single-phase) liquid.

Before completing this section, we can use the microstructure approach to understand the most basic behaviors and properties of tool steels, as shown in Figure 3.3. For example, plastic molding steels need high resistance at room temperature, but not at high temperature, and usually do not require wear resistance through added particles. Actually, the absence of hard particles is often desirable, since hard particles lead to polishing defects (e.g., those called pin holes). The same absence of particles also occurs in hot work tool steels (and this is primarily the reason why hot work steels are often used in plastic molds), not because of surface appearance but because of the detrimental effect of hard particles on the toughness. On the other hand, the matrix of a hot work tool steel should have high stability at high temperatures because the surface of the tool is heated (usually above 500°C) when forming parts in hot-forming processes. The same does not happen in plastic molds, where the surface is maintained close to room temperature, or even colder, by cooling channels. In cold work and high-speed steels, the presence of particles is important because both applications require high wear resistance. This is different in the case of machining tools, where high-speed steel is mainly used, because the tool's surface gets heated as a result of machining friction and the cutting operation itself. So the matrix of a high-speed steel must have stability at higher temperatures, while this is not necessary for cold work tool steels.

	Mold Steels	Cold Work Steels	Hot Work Steels	High-Speed Steels
Matrix — Room temperature strength (hardness)	+	++	+	++
Hot resistance (precipitation hard.)	–	–	+	+
Toughness (at working hardness)	+	–	++	–
Particles — Wear resistance through carbides	–	++	–	++
Clean microstructure, in terms of inclusions or carbides	++	–	+	–

FIGURE 3.3 Properties of tool steels for different applications: "+" means that the property is important, "++" means very important, and "–" means that usually this property is less critical than the others.

In conclusion, the above described "microstructure approach" enables a much better understanding and reduces substantially the complexity in understanding tool steels. In this book, several examples of microstructures are given, and the correlation of the final properties will be clear to the reader. Then, whether or not one is observing an actual tool steel under a microscope, the microstructure can always be referred to for enabling an understanding of either the properties or performance by the undissolved particles or the microstructural matrix. Usually, optical microscopy is enough to describe most of the phenomena and is primarily used along the present book.

3.2 HEAT TREATMENT OF TOOL STEELS

The intention of heat treatment is to change the characteristics of the matrix. Small carbides are dissolved, but the primary or eutectic carbides (the particles described earlier) will suffer only minor changes or will not change at all after heat treatment. The matrix-related properties described in Figure 3.3 are then deeply affected by the heat treatment, such as low- and high-temperature strength, toughness, or other mechanical properties. So the basic steps of heat treatment are discussed in this section.

If not the most important, heat treatment of tool steels is one of the most critical points in the use of these steels, since the properties of tool steels are achieved only after the final heat treatment. The change in mechanical properties during heat treatment is very dramatic, increasing, for example, the hardness from 20 HRC to more than 60 HRC, which is equivalent to an increase of more than three times in tensile strength. This is possible only because of the strong ability of tool steels to increase the hardness after quenching and, in most cases, to achieve a second step of hardening during tempering.

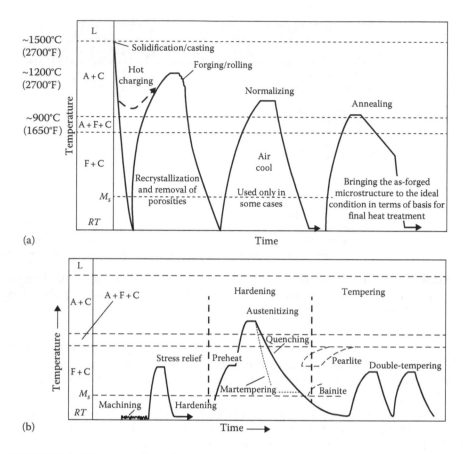

FIGURE 3.4 Phase transformations during the thermal history of tool steels, split in two different groups. (a) Heat treatment performed by the steel mill, during the production of the tool steels. (b) Final heat treatment, usually done after the tools are rough-machined by heat treatment companies. L, liquid; A, austenite; C, carbides; F, ferrite, M_s, martensite start temperature. (Modified from Branco, J.R.T. and Krauss, G., Heat treatment and microstructure of tool steels for molds and dies, *Proceedings from Tool Materials for Molds and Dies*, Krauss, G. and Nordberg, H., eds., Colorado School of Mines Press, Golden, CO, 1987, pp. 94–117.)

By definition, all steps that involve changes of microstructure and properties after a thermal cycle can be considered as heat treatment. However, for tool steels, they are clearly split into two major groups (Figure 3.4). The first group relates to the heat treatment and phase transformations that are effected during steel production (Figure 3.4a). Right after solidification, there are several steps during which the microstructure changes from austenite to ferrite (or bainite, or even martensite), which will affect the final microstructural condition and therefore should be considered. The last heat treatment done by the steel mill is usually annealing, which leads to low hardness and proper condition for the machining and the final heat treatment of a given tool. After tool machining and before putting them into operation, tools are subjected to the final heat treatment, which is usually hardening and tempering (Figure 3.4b). This treatment is the one that is more closely related to the final properties and has strong relevance to the tool's performance. As seen in the upcoming sections, both groups of heat treatments are considered, but greater focus is given to the second group, because it is more complex and has a higher effect on the final properties.

3.2.1 HEAT TREATMENTS PERFORMED BY THE STEEL MILL

If one asks which heat treatments are applied by the mill before delivering a tool steel block or bar, the immediate response would be "annealing." While this is correct because the annealing treatment is usually the last one done by the steel mill, all the thermal history during tool steel production has its importance as well, starting from the moment the final composition is cast into ingots.

From the moment of *solidification*, the first solid structure (grains) is formed, and the first phase transformation may occur (see first cooling line of Figure 3.4a). It is a common practice by steel mills that the ingots are loaded in the reheating furnace in the hot condition (hot charging), usually above 700°C, thus before the transformation from austenite to ferrite (martensite or bainite) has started. The first reason for this practice is to save energy and time in the reheating of ingots during forging or rolling, but it is also used to avoid casting cracks that occur after solidification and the cooling down to room temperature in the ingot molds. This likelihood of the emergence of casting cracks exists because most tool steels have high hardenability, due to the high content of carbon and alloying elements; therefore, the solidification inside the ingot molds may be fast enough to cause the transformation of high-hardness constituents (martensite or bainite) in a coarse microstructure. As a result, this dramatically reduces the toughness and leads to cracking due to thermal and transformation stresses.

After solidification and reheating, but before hot-forging or rolling, the microstructure is very coarse because of the inheritance of the as-cast structure and also because of the grain growth during the long times and high temperatures of reheating. The forging or rolling will then break down the initial as-cast microstructure (as discussed in Chapter 2 and further in this chapter, in Section 3.3.3.2), leading to a homogeneous distribution of carbides and refinement of the grains. However, an important acknowledgement has to be made on tool steel regarding this last point. Most metallurgical books explain that the hot-forming operation is responsible for refinement of the microstructure. While this is undeniably important for tool steels, there are other factors that play very important roles in the grain refinement of tool steels, as can be seen from the following paragraphs.

The first important peculiarity of tool steels is the amount of deformation: while for rolled bars it often exceeds 1000 times (on an area basis), the same is not possible in large forged blocks. The deformation applied is usually calculated to close the internal voids in the ingots, and is in the range of four to eight times in area, meaning that it is not sufficient to enable strong microstructure refinement. Consequently, the as-forged austenite grain of a large ingot is in the range of 0–2 ASTM. On the other

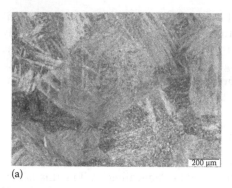 

(a)   (b)

FIGURE 3.5 Example of microstructure of a large block of modified P20 (section size of ~900 mm), from the center of the block, under same magnification. (a) Cooling down after forging. (b) Normal condition, after annealing and quenching. Both microstructures are primarily bainitic, but the prior-austenite is very different: about 0 ASTM for as-forged and 5 ASTM for the annealed and hardened. (From Bacalhau, J.B., Development of a 40 HRC plastic mold steel with high machinability, Master thesis of the Graduate Course in Aeronautic and Mechanical Engineering, Instituto Tecnológico de Aeronáutica, São José dos Campos, Brazil, 2012, 140pp., Original title: *"Desenvolvimento de aço para moldes plásticos com 40 HRC e elevada usinabilidade."*)

hand, the final prior-austenite* grain size of a hardened block of H13 or P20 is often in the range of 5–10 ASTM. (See the example in Figure 3.5.) The difference is caused by the further heat treatment after forming, especially by the *annealing treatment*, performed after forging and prior to the final heat treatment hardening, as detailed next.

If a tool steel is evaluated immediately after hot forging or rolling, the microstructure is composed of mixed phases, usually martensite and bainite, which is transformed from a coarse prior-austenite grain size and presents very high hardness. For similar reasons that ingots are not cooled down to room temperature (saving energy and the risk of cracks), forged blocks and large rolled bars are usually also hot-charged to the final annealing treatment. During this treatment, the microstructure, regardless of whether it is austenite, martensite, or bainite, is converted isothermally or during cooling to ferrite and carbides. Carbides are typically alloy carbides, rich in Cr, Mo, W, or V, and therefore are spherical and not lamellar-shaped such as perlite in carbon steels (see Figure 3.6 for examples of proper and improper annealing conditions). Compared to the as-cast condition, one can easily identify hundreds of carbide particles in a 500× magnification microstructure; those particles will dissolve during austenitizing, but before dissolution will serve as nucleation sites for austenite grains and in turn will generate a large number of austenite grains. Then, considering that the final heat treatment does not exceed in time and temperature, the austenite grain size will be much finer than the as-forged grain size. This explains

---

* The term "prior-austenite grain size" refers to the austenite grains that existed during the hardening (or austenitizing) and were transformed to martensite or bainite, during cooling. This is carefully defined in Section 3.2.2.

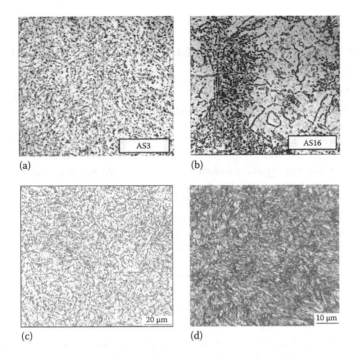

FIGURE 3.6 Example of annealed microstructures on H13 tool steel. (a) Acceptable and (b) unacceptable according to NADCA (500× magnification). (c) Example of a large-diameter rolled bar, where the refined and annealed microstructure is converted to (d) a final homogeneous microstructure after hardening and tempering. (a and b: According to NADCA #229/2006, Special quality die steel & heat treatment acceptance criteria for die casting dies, North American Die Casting Association—NADCA, Holbrook Wheeling, IL, 2006, 33pp.; Pictures (c) and (d): From Silva et al., Implementation of a rough rolling for high alloy steels at villares metals S.A., 2004, *Proceedings of 59 Congresso Anual da ABM (59th Annual Conference of Brazilian Metallurgical and Materials Society)*, São Paulo, Brazil, 2004, pp. 297–305.)

then why the as-forged microstructure in Figure 3.5 is much coarser than that after annealing and reaustenitizing.

To obtain this adequate annealing microstructure, two main types of annealing cycles can be used, with different characteristics, discussed next and summarized in Table 3.1. The first possibility is when the annealing treatment consists of full austenitizing, then named *full annealing*, followed by slow cooling, when carbon and alloy elements will slowly precipitate as spherical carbide particles. Common temperatures and cooling rates for full annealing are given in Table 3.2. These carbides are named *secondary carbides*, because they are formed by a solid-state reaction and not during solidification.* The final metal matrix will then be composed of

---

* In Section 3.3, we will explain how the solidifying liquid undergoes several reactions, to form either primary or eutectic carbides, much larger than the secondary carbides.

TABLE 3.1

Schematic Comparison of General Characteristics of Subcritical and Full Annealing for Tool Steels

Subcritical Annealing	Full Annealing
Faster with finer microstructure, but higher hardness than full annealing.	Low hardness but a longer treatment and with coarser distribution of alloy carbides.
Not sensitive to decarburizing.	Decarburizing can occur if furnace atmosphere is not controlled.
Requires high hardenability or previous austenitizing and quenching, if used as final annealing. Also requires cooling down room temperatures below M_s (300°C or lower) after hot forming, before annealing starts.	Hot charging can be used, directly after forging or rolling.
Critical for high-speed steels, due to high hardness, and for some cold work steels, because of cracking after cooling down from hot forming operations.	Critical for some hot work and plastic mold steels, due to coarse carbides precipitated along grain boundary during slow cooling.
Usually applied for semis and during final annealing of high-alloy hot work and some plastic mold steels.	Extensively applied for most tool steels.

ferrite, with low carbon and low amount of alloy elements in solid solution, and the dispersion of fine secondary carbides. Therefore, the lowest final hardness is usually obtained after full annealing.

An alternative process for annealing is called *subcritical annealing*. It consists of heating the as-forged or as-rolled bars to temperatures below the austenite formation (below the A1 temperature) so that the same carbides can be formed by precipitation from the initial microstructure (martensite or bainite). The treatment is thus similar to a tempering treatment and sometimes referred to in the literature as *temper annealing*. The first advantage of subcritical annealing is that it usually leads to a more refined microstructure after final hardening and tempering, due to the fine and homogeneous dispersion of secondary carbides in the annealed state, which promotes intense nucleation during austenitizing. However, those smaller carbides also lead to a higher hardness in the annealed condition, which is one of the biggest disadvantages of subcritical annealing. Another advantage of subcritical annealing is that it can be done, usually, in shorter times and does not cause extensive decarburizing, since austenite is not formed. So it is a very convenient annealing cycle for semifinished products in the steel mill, such as billets or intermediate forged products. On the other hand, subcritical annealing requires a homogeneous microstructure after bars are cooled down from rolling or forging to room temperature or at least ~200°C. This may occur naturally by air-cooling of small-diameter (<50 mm) rolled bars of high-alloy tool steels (e.g., high-speed steels), but not in other kinds of tool steels. So, if subcritical annealing is applied as the final annealing cycle in hot work or plastic mold steels, a previous austenitizing and quenching cycle is recommended to condition the initial microstructure, usually with oil quench for large bars (>200 mm) or air-blast for smaller bars. This previous quenching may be critical for cold work steels, in terms of cracking, and subcritical annealing is not common in this class.

TABLE 3.2
Example of Most Annealing Temperatures (for Full Annealing) for Most Common Tool Steels

ASTM	DIN	Critical Temperatures (°C)			Full Annealing	
		Ac_{1b}	Ac_{1e}	M_s	Temperature	Hardness (HB)
Hot work tool steels						
H10	1.2365	810	890	280	870°C–900°C/1600°F–1650°F	<229
H10 mod	1.2367	820	890	300	870°C–900°C/1600°F–1650°F	<229
H11, Low Si H11, H13	1.2343, 1.2344	830	900	340	870°C–900°C/1600°F–1650°F	<229
6F3 mod	1.2714	720	790	260	760°C–790°C/1400°F–1450°F	<212
Cold work tool steels						
A2	1.2363	780	820	230	845°C–870°C/1550°F–1600°F	<229
W1	1.2025	720	800	180	750°C–790°C/1380°F–1450°F	<201
D2	1.2379	830	860	200	870°C–900°C/1600°F–1650°F	<255
D3 or D6	1.2436	790	820	210	870°C–900°C/1600°F–1650°F	<255
8%Cr	—	850	900	260	870°C–900°C/1600°F–1650°F	<255
O1	1.1510	730	770	150	760°C–790°C/1400°F–1450°F	<212
S1	1.2550	780	860	300	790°C–820°C/1400°F–1500°F	<229
S7	1.2550	820	860	280	815°C–845°C/1500°F–1550°F	<229
Plastic mold steels						
P20 mod	1.2738	720	810	300	760°C–790°C/1400°F–1450°F	<179
420 mod	1.2083	820	900	280	800°C–850°C/1470°F–1560°F	<229
High-speed steels						
M2	1.3343	810	840	150	870°C–900°C/1600°F–1650°F	<241
M42	1.3247	840	860	150	760°C–790°C/1400°F–1450°F	<269

Sources: Data from Roberts, G. et al., *Tool Steels*, 5th ed., ASM International, Materials Park, OH, 1998, pp. 75–77; Chandler, H., *Tool Steels, Heat Treaters Guide: Practice and Procedures for Irons and Steels*, ASM International, Materials Park, OH, 1995, pp. 517–669; Ohlich, J. et al., *Atlas zur Wärmebehandlung der Stähle, Bd. 3, ZTA-Schaubilder*, Verlag Stahleisen mbH, Düsseldorf, Germany, 1973; Uddeholm Tooling, Technical catalogue for tool steels, Available online at: http://www.uddeholm.com/files/, Accessed December 2015; Böhler Edelstahl, Technical catalogue for tool steels, Available online at: http://www.bohler-edelstahl.com/english/1853_ENG_HTML.php, Accessed December 2015; Villares Metals, Technical catalogue for tool steels, Available online at: http://www.villaresmetals.com.br/pt/Produtos/Acos-Ferramenta, Accessed December 2015.

Note: The AC_{1b} temperature, which is the maximum temperature before austenite is formed, is also shown as a reference for subcritical annealing.

In addition, for both cold work and high-speed steels, the high amount of carbon and undissolved carbides (particles, see Section 3.3) naturally increases the annealed hardness, such that the annealing cycle leads to the lowest possible hardness and then the full annealing is commonly applied. In terms of mechanism, like a tempering treatment, the higher the subcritical temperature, the lower the hardness. As shown in Table 3.2, the AC_{1b} temperature is also shown for the most common grades, representing the initial temperature of austenite formation and thus the theoretical maximum limit of subcritical annealing. In practical terms, because of temperature (furnace) or composition (segregation) variations, the maximum temperature for subcritical annealing is between 30°C and 50°C (50°F to 90°F) below the AC_{1b} temperature.

Before ending this section, one aspect is also important to consider before obtaining a homogeneous annealing microstructure. After solidification, there is a strong tendency for segregation in tool steels, meaning that the as-cast microstructure does not present carbon and the alloying elements homogeneously distributed. The topic of segregation is outside the scope of this book (see Reference 6), but, in short, two main points have to be observed to reduce segregation: (1) the solidification parameters, such as superheat, ingot size and geometry, and other casting variables (see example in Figure 3.7a), and (2) the reheating conditions prior to forging or rolling, using long times (several hours) at very high temperatures (usually 50°C–100°C form the solidus temperature), which tends to allow the diffusion of alloying elements within the interdendritic spacing, leading to a more homogeneous composition in the final microstructure (Figure 3.7b). This treatment is often called *homogenizing*. Observe in Figure 3.6b that the example of bad (not approved by NADCA [4]) annealing condition is clearly related to the inhomogeneous population of secondary carbides, which in turn are determined by the annealing cycle applied and also by an inhomogeneous distribution of alloying elements, or segregation. While it is common to evaluate the annealed condition only by the hardness, those examples show how the microstructure of heat-annealed steels are also important and may substantially vary depending on the annealing cycle and segregation.

In summary, obtaining the desired microstructure after the final heat treatment also depends on the initial microstructure: the annealed state. And this microstructure will result from proper processing (solidification and forming), in addition to the thermal history of the produced tool steel, which includes the annealing heat-treatment cycle.

3.2.2 FINAL HEAT TREATMENT OF SEMIFINISHED TOOLS

Once a tool steel block or bar is properly produced by the steel mill with the expected composition, hardness, and microstructure from the annealing treatment, the manufacturing of the final tool can start. Typically, rough machining is the first step. Most of the machined volume is removed during this phase by metal cutting operations, mainly turning or milling. In some tools, such as plastic molds, die-casting tools, and sheet forming dies, the volume of material removed by machining is extremely high, often reaching more than 50% of the volume of the original piece. So, among other characteristics, it is usually the desired reduced hardness in this manufacturing step that enables fast cutting by machining. For this reason and also from metallurgical aspects

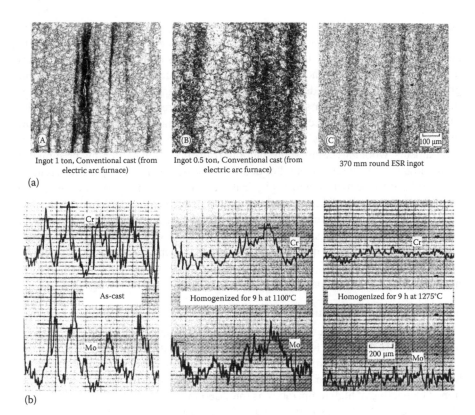

(a)

(b)

FIGURE 3.7 Segregation examples in hot work steels. (a) Effect of ingot type on the segregation in 75 mm rolled bars of a modified H10 tool steel (DIN 1.2365). (b) Microsegregation in the central regions of a 600 mm diameter ESR ingot, subjected to different homogenizing times. The variations in alloying elements was measured by microprobe analysis. All data are for H13 tool steel. (From Wilmes, S. and Burns, K.P., *Gießerei*, 76(24), 835, 1989.)

(such as quenching restrictions; see Section 3.2.2.2), the final hardening heat treatment is often performed after the final machining. After the tools have been heat-treated, they are machined to the final dimensions, removing stock allowance of a few millimeters* typically left from the rough machining. Final machining, by milling, turning, grinding, or EDM (electrodischarge machining), is applied to remove dimensional changes and distortions from heat treatment, as well as possible surface decarburizing.

The final heat treatment, from here on referred to as simply *heat treatment*, is responsible for bringing the tool steels to their final properties, which are dramatically different from those of the annealed state. In fact, one of the common characteristics of tool steels is the interdependence of properties and performance on the heat treatment. Induced strain by machining (surface work-hardening) or residual stresses may be eliminated by the recrystallization and phase transformation that

---

* The allowances depend on the steel type, the quenching process, and the geometry, but are generally more than 0.3% of the dimension, for example, 1.5 mm for a 500 mm tool.

occur during the heating phase to hardening. During austenitizing, several alloy carbides are dissolved in austenite, affecting the solid solution of the upcoming constituents (martensite or bainite), which are intrinsically hard due to the phase transformation during quenching. These final constituents also present much finer grains (often named *laths*, due to their morphology), and the alloy elements present in the solid solution of their structures may promote precipitation during tempering. This description is very succinct for all the effects in final heat treatment, but shows that heat treatment affects and substantially changes all possible strengthening mechanisms: work hardening, solid solution, transformation hardening, precipitation hardening, and grain refinement. For details on those mechanisms, see Reference 1.

In addition to the many complex physical metallurgy reactions, tool steel heat treatment deals with many different steel compositions as well as a diverse range of furnaces and other equipment used for heat treatment. However, the understanding of tool steel heat treatment can be simpler once it is divided in two main phases*: the hardening (with austenitizing and quenching steps) and tempering, as schematically depicted in Figure 3.4. This is the objective of the present section so that at the end of it the full mechanism and the effect on properties and performance are easily understood.

3.2.2.1 Hardening I: Austenitizing

"Hardening" is a general term for metallic materials and can be used for any kind of increase in strength, such as solid solution or precipitation hardening. However, in tool heat treatment, hardening means the rapid cooling of a given steel block from high temperature to obtain high hardness, usually through martensite transformation. The heating at high temperature is intended to form austenite, the high-temperature phase of tool steels, so the first part of hardening is the austenitizing treatment. The fast cooling is more scientifically referred to as *quenching*, which may include several different facilities and techniques.

Before starting the austenitizing treatment, some initial considerations should be addressed. First, the rough machined tool should be considered, especially which kind of machining technique was used and to what extent it was applied. If the machined volume is high (>30%–50% of the initial block), a *stress relief treatment* is recommended, which consists of slow heating (less than 100°C/h) to temperatures between 500°C and 600°C, followed by cooling in still air. The treatment time is a minimum of 2 h after the entire piece has reached the treatment temperature, and in practice it is recommended 0.5 h/in. of the largest section, preserving a minimum of 2 h treatment regardless of the tool size. In addition to reduce distortion and cracking during the final heat treatment, the stress relief treatment gives the chance that thermal dimensional changes are adjusted by machining, before the full hardening is applied. Stress relief treatment is also suggested in situations where the tool is subjected to substantial mechanical or thermal stresses, such as after tooling operations[†] or welding. If applied to tools already heat-treated by hardening and tempering, the

---

* For tools from which a large volume of the initial tool steel block is removed by machining, a stress-relief treatment should be applied before austenitizing.

[†] To improve tools performance, some companies have the practice of applying a stress-relief heat treatment to used tools before putting them back into operation, or even on a regular basis after a certain production run.

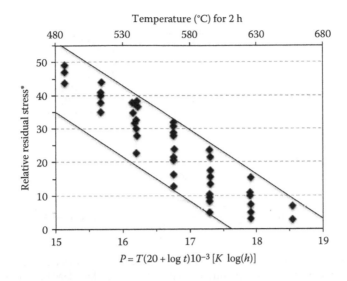

Temperature (°C) for 2 h

$$P = T(20 + \log t)10^{-3} \, [K \, \log(h)]$$

FIGURE 3.8 Diagram showing the reduction of residual stress with stress relief treatment. The relative residual stress means the amount of stress divided by the steel's room temperature yield strength. Data is plotted versus the Larson–Miller parameter, which correlates the time (t) and temperature (T), with time given in hours and temperature in kelvin. *Relative residual stress means the remaining stress divided by the steel room yield strength, in percentage. (Drawn with data from Rosenstein, A.H., *J. Mater.*, 6(2), 265, 1971.)

recommended stress relief temperature is ~50°C less than the (highest) temperature of the original tempering. As a reference, Figure 3.8 exemplifies how the stresses are reduced with increase in time and temperature.

Either after stress relief or in cases where stress relief is not necessary, hardening of tools requires *austenitizing treatment*. In this step, heating to the required temperature should be controlled because fast heating may cause distortion or even cracks. The reason is that fast heating leads to a temperature difference from the surface to the core (center) of a tool, marked as ΔT in Figure 3.9. One practical way to reduce ΔT is to introduce one or more preheating phases in which the furnace is maintained at a given temperature to allow temperature homogenizing. Usually, two preheating steps are employed, one before the austenite formation and one after, typically at 600°C–650°C and 800°C–850°C (1110°F–1200°F and 1470°F–1560°F, respectively), respectively. For salt-bath furnaces and heat treatment of high-speed steels with very high austenitizing temperatures, another preheating step is introduced at ~1100°C (2010°F).

In tool steels, the austenitizing temperatures are usually well above the common range of 750°C–850°C (1380°F–1560°F) used for carbon steels. The reason is that in (hypoeutectoid) carbon steels, austenitizing has the unique purpose of forming austenite, where the cementite (iron carbides) is quickly dissolved, so it is a common practice to austenitize carbon steels 50°C above the A3 temperature, where the microstructure will be fully austenitic. In tool steels, all the ferrite must be eliminated just as in carbon steels. The difference is that in tool steels there are usually

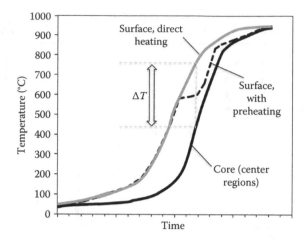

FIGURE 3.9 Schematic heating behavior of the surface and core of a large tool, showing the difference in temperature from the surface to the core (ΔT), which leads to thermal stresses. The introduction of a preheating step, here simulated at 600°C and 850°C, leads to lower ΔT values and then to a reduction in residual stresses.

secondary or primary carbides that do not dissolve immediately after the matrix is completely converted from ferrite to austenite. On the contrary, those carbides gradually dissolve as the temperature increases, bringing more carbon and alloying elements into solid solution in austenite. The reason for the difference is the high stability of alloy carbides versus cementite (Fe_3C), especially for the Cr, Mo, V, and W carbides, which are the most traditional alloy elements in tool steels (see Figure 3.10).

To express this mechanism of increasing the hardening temperature and the gradual dissolution of alloy carbides, Figure 3.11 can be used for describing the solubility curves for the most common alloying elements. In the case of Cr and Mo,* the data are estimated, but they are sufficient for the purpose of the examples given here. Observe that curves in the upper region indicate higher solubility, thus illustrating that for a given temperature, a higher amount of elements is in solution. For example, let us consider a 0.22 wt.% C, which would represent a 1 at.% C in a hypothetical C–Fe–M alloy, where M is the alloying element. For a fixed hardening temperature of 1000°C, it is easy to determine the maximum amount of alloying elements that can be added to the steel and still be placed in solid solution. For this example, running the conversion from at.% to wt.%, the maximum amounts of soluble alloying elements that could be added are ~8 wt.% Cr, 2.5 wt% Mo, 0.5 wt.% V, and less than 0.01 wt.% Nb or Ti. This shows that by means of dissolution and further precipitation during tempering[†] Cr, Mo, and V are the most appropriate elements, and in fact several tool steel compositions show high amounts of those elements for this purpose. However, for higher carbon contents and for usual hardening temperatures, part of those carbides may not

---

* W carbides are similar as Mo, in terms of solubility product in atomic percent.

[†] This precipitation causes the phenomenon of *secondary hardening*, discussed in detail in Section 3.2.2.3.

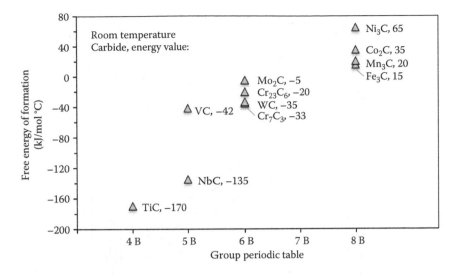

FIGURE 3.10 Free energy of formation of carbides for carbide systems in iron at room temperature, for various alloy elements. All data are for 1% activity of the metal, which by means of simplification can be considered as 1% content. The data represents the free energy of formation per mole of carbon, and thus also called the *carbon potential*, leading to the decrease of free energy when 1 mol of carbon is added and forms the given carbide in the alloy. Values were extracted from carbon potential versus temperature diagrams [14,15], which are similar to the oxygen potential diagrams (Ellingham diagrams): the lower the energy, the higher the carbon's tendency to form carbides.

be fully dissolved. Again, in the same example, by increasing the carbon to 1 wt.% (almost 5 at.% C) and running the same calculations, we can determine that the limit of solubility at 1000°C would be five times lower: ~1.6 wt.% Cr, 0.5 wt.% Mo, 0.1 wt.% V, and much less than 0.005 wt.% Nb or Ti. In the case where a steel composition contains higher amounts of alloying elements than the values given earlier, the excess alloy contents will stay combined with carbon, forming undissolved carbides. In several applications, as detailed in Section 3.3, undissolved carbides are necessary and the hardening is not supposed to dissolve all those carbides. In other situations, however, a higher dissolution of carbides may be necessary to achieve higher hardness after tempering, and therefore the hardening temperature should be increased.

Therefore, tool steels' austenitizing temperatures vary considerably for different steel types. Table 3.3 shows the usual temperatures for the most common grades. Those temperatures are also found in the brochures of tool steel producers,* and should be followed. If a temperature lower than the specified is applied, the solubility of the alloy carbides may not be sufficient. For example, Figure 3.12 shows an example of the dissolution of carbides in various high-speed steels, indicating how the volume fraction of carbides gradually decreases with the increase in hardening

---

* The author advises that the producer's temperature values be considered in case of doubt, since the compositions may vary for the same grade produced by different steel mills.

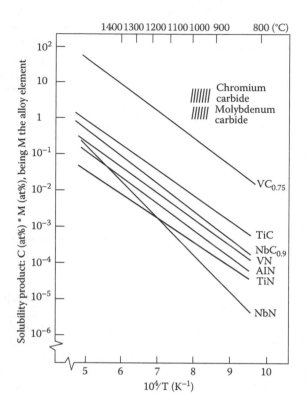

FIGURE 3.11 Solubility of alloy carbides in austenite. The solubility lines above represent carbides that require less austenitizing temperatures to go into solution (see text for explanation). (From Porter, D.A. and Easterling, K.E., *Phase Transformations in Metals and Alloys*, CRC Press/Taylor & Francis, Boca Raton, FL, 2004.)

temperature. By being more stable (see Figure 3.10), MC carbides dissolve at a lower rate than the other carbides.

On the other hand, when the temperature is too high, several problems may occur, including "metallurgical damage" to the tool steel and a decrease in its properties, excessive dimension changes or distortion to the tools, and also damage to the heat-treatment equipment. Focusing only on the aspects of physical metallurgy, three main problems may occur when the temperature is too high:

1. Excessive grain growth of austenite, leading to coarse microstructure and poor toughness. For reference, Figure 3.13a through c show several examples of grain size variation with austenitizing temperature.
2. Local formation of liquid, especially at the micro-segregated areas, which are richer in carbon and alloying elements and present lower solidus temperature. See example in Figure 3.13c and d.
3. Excessive dissolution of alloying elements and carbides. First, this may lead to quench embrittlement, caused by the strong tendency of proeutectoid carbides on grain boundaries during quenching. Quench embrittlement is

shown in detail in Sections 3.2.2.2 and 3.2.2.3, also demonstrating how this gets stronger when higher austenitizing temperatures are used. And second, the increase in dissolution may decrease the martensite start and finish temperatures, which then increase the amount of retained austenite (see Section 3.2.2.2 for a detailed explanation). The large amount of this phase, in turn, may lead to lower hardness and unexpected embrittlement after tempering or dimension instability during tool use, in this last case when austenite transforms to martensite due to tooling stresses.

Similar to the temperature, the austenitizing time also should be controlled, because of the effects of longer and shorter times similar to higher or lower temperatures. However, the time effect is usually far less strong than the temperature effect. As a general rule, the austenitizing time is 30 min at the temperature for all tool steels, with the exception of high-speed steels, when temperatures are very close to solidus and times are only between 2 and 5 min. Ideally, the time should be controlled by thermocouples inserted into holes present in the actual tools (which is rather common, such as cooling channels in plastic mold or die-casting dies) or in dummy blocks with similar dimensions. In case both are not possible, the rule of 30 min for each inch is also applied, but this is rather imprecise and a better understanding may be important.

3.2.2.2 Hardening II: Quenching

After full austenitizing, meaning that all microstructural matrix has converted into austenite and that the secondary carbides are dissolved, the tool steel is ready to be quenched. Quenching in tool steels follows the same basic principles as in carbon and low-alloy steels, meaning that it aims to produce martensite by fast cooling from the austenite phase. And martensite transformation also follows the same mechanism of coordinated movement of dislocations, leading to fast and diffusionless transformation from the face-centered cubic austenite to body-centered tetragonal structure. However, there are several specific points in quenching of tool steels that need to be addressed, since they dramatically affect the final properties and quality of a given tool. These points may be divided into three main categories: (1) the large volume and hardenability considerations in tool steels and the likelihood for distortion; (2) the precipitation of carbides on grain boundaries, also known as *quench embattlement*; and (3) the formed microconstituents after quenching, in terms of balance between ferrite, pearlite, martensite, bainite, and retained austenite.

For *volume effects, hardenability*, and *distortion*, the discussion can start by considering that tool steels usually show much higher hardenability than carbon steels, meaning that the necessary cooling rate to bypass the pearlite field is not typically high. Therefore, it is common that tool steels are quenched in oil or high-pressure nitrogen, rather than the water quench applied to carbon steels. For example, Figure 3.14 shows the difference in the critical cooling rate between a 8%Cr high-carbon tool steel and a low-alloy steel—where the high amount of C and alloying elements pushes the ferrite/pearlite transformation to longer times, in this case from 2 s in the low-alloy steel to more than 1000 s in the high-alloy tool steels.

TABLE 3.3
Hardness after Tempering for the Most Common Tool Steels, Considering Hardening in Usual Temperatures

ASTM	DIN	Hardening	As Quench	Hardness (HRC) Tempering, 2 × 2 h						
				200°C (390°F)	250°C (570°F)	500°C (930°F)	525°C (980°F)	550°C (1020°F)	600°C (1110°F)	625°C (1160°F)
Hot work tool steels										
6F3 mod	1.2714	900°C (1650°F)	59	55	53	45	44	43	41	38
H10	1.2365	1020°C (1870°F)	52	51	50	52	52	52	50	48
H10 mod	1.2367	1030°C (1885°F)	56	54	53	54	54	53	50	46
H11	1.2343	1010°C (1850°F)	55	54	53	55	54	53	46	41
Low Si H11	1.2343 mod	1010°C (1850°F)	53	53	52	52	52	51	46	42
H13	1.2344	1020°C (1870°F)	55	54	53	55	55	53	47	42
H21	1.2581	1100°C (2010°F)	50	49	49	51	52	51	50	48
Cold work tool steels										
A2	1.2363	950°C (1640°F)	64	62	60	58	57	54	47	42
W1	1.2025	800°C (1470°F)	64	62	59	na	na	na	na	na
D2	1.2379	1020°C (1870°F)	64	63	59	59	57	54	45	40
D3 or D6	1.2436	960°C (1750°F)	65	61	60	56	53	51	44	40
8%Cr	—	1030°C (1885°F)	65	62	60	62	61	59	49	45
O1	1.1510	820°C (1505°F)	64	62	60	46	45	42	39	36
S1	1.2550	925°C (1700°F)	59	56	55	46	45	44	42	na
S7	1.2550	950°C (1640°F)	60	57	55	52	48	45	42	na

(Continued)

TABLE 3.3 (*Continued*)

Hardness after Tempering for the Most Common Tool Steels, Considering Hardening in Usual Temperatures

| | | | | | | Hardness (HRC) | | | | | |
| | | | | | | Tempering, 2 × 2 h | | | | | |
ASTM	DIN	Hardening	As Quench	200°C (390°F)	250°C (570°F)	500°C (930°F)	525°C (980°F)	550°C (1020°F)	600°C (1110°F)	625°C (1160°F)
Plastic mold steels										
P20 mod	1.2738	850°C (1560°F)	52	51	50	42	40	37	34	33
420 mod	1.2083	1030°C (1885°F)	56	53	52	54	53	43	37	35
High-speed steels										
T1	1.3355	1260°C (2300°F)	64	61	62	64	65	65	62	60
M2	1.3343	1200°C (2190°F)	65	61	62	64	65	65	62	60
M2	1.3343	1100°C (2010°F)	63	61	60	62	62	61	58	54
M42	1.3247	1180°C (2155°F)	64	61	60	69	68	67	63	61
M50	—	1100°C (2010°F)	63	61	60	62	63	63	58	55

Note: Important! The exact values after heat treatment depend on the alloy design of each tool steel manufacturer, which can vary substantially within the ranges of a given grade. The following data were obtained from brochures of traditional tool steel producers [10–12]. "na" means not available.

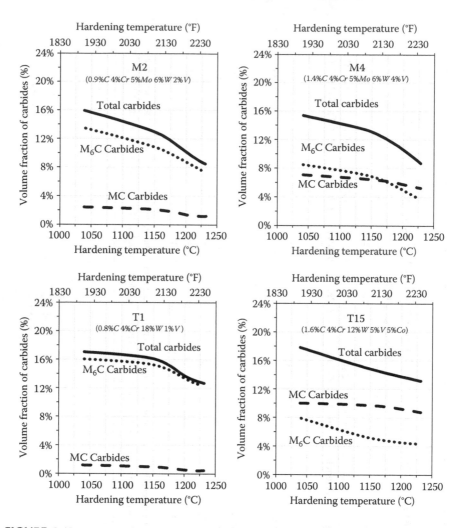

FIGURE 3.12 Volume fraction of undissolved carbides in eight different types of high-speed steels. Observe that the overall carbide volume fraction decreases when the temperature increases, meaning that carbides gradually dissolve when hardening temperature increases. Also notice that the MC-type carbides (rich in V or Nb) are more stable and dissolve to a lesser extent than M_6C carbides (rich in Mo or W). (Adapted from Kayser, F. and Cohen, M., *Met. Prog.*, 61(6), 79, 1952.)

This improvement in hardenability is directly related to the amount of carbon and alloying elements in solid solution after austenitizing, being strictly important to guarantee hardening throughout in large tools, such as plastic molds or large forging dies, with section size up to 1000 mm (40 in.). To quantify this effect of dimensions, the curves in Figure 3.15 show the cooling profile of core regions in bars of different sizes when cooling in water, oil, and air. Taking the example of oil cooling, it can be seen that large diameters (such as 500 mm) need ferrite to be pushed to more than 20 min (or >1000 s) in order to avoid the formation of this constituent in the core

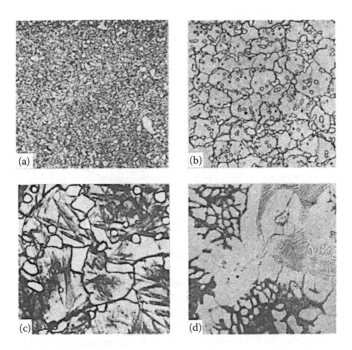

FIGURE 3.13 Examples of austenitizing temperatures in a T1 high-speed steel. (a) 870°C (1600°F), typically the initial of austenitizing formation, but with insufficient dissolution of carbides. (b) 1260°C (2300°F), a typical austenitizing temperature at which the secondary carbides are dissolved and also part of the primary/eutectic carbides. In (c) 1290°C (2350°F), the high-speed steel shows clear signs of overheating, with much larger grain size and carbide coarsening by coalescence, with the carbides changing their shapes to square (instead of round), meaning that the temperature is high enough so that the most favorable crystallographic plans can be developed. This microstructure does not promote advantages in terms of hardening, but only loss in toughness. (d) 1320°C (2410°F), at which typical signs of melting are noticed and the carbides assume again the as-cast condition (dendritic-like). This is often called *burning* and makes the high-speed steel to be very brittle and inadequate for use condition. In (c) and (d), the amount of retained austenite also increases, decreasing the hardness and increasing the instability of the final tool. (All pictures from Roberts, G.A. and Cary, R.A., *Tool Steels*, 4th edn., American Society for Metals, Materials Park, OH, 1980, pp. 645–653.)

regions. Therefore, in the example of the two steels from Figure 3.14, the low alloy steel would clearly not achieve full martensite transformation, while the high alloy tool steel would. So, the hardenability of tool steels is high but also the dimensions of tools are larger when compared to heat-treated components of low-alloy or low-carbon steels.

A good example of this dependence of hardenability and dimensions is the popular 1.2738 and 1.2311 plastic mold grades, which are very similar, but 1.2738 has about 1% Ni to improve the hardenability and enable the hardening of sections above 500 mm, while 1.2311 is usually restricted to plastic mold plates less than 400 mm thick.

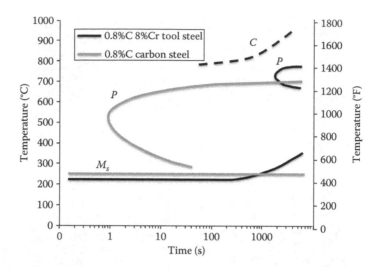

FIGURE 3.14 Continuous cooling transformation (CCT) diagram for a 0.8%C carbon steel and for a high-alloy tool steel (0.8%C, 8%Cr, 2%Mo, 0.5%V). P represents the pearlite transformation, M martensite, and C the precipitation of proeutectoid carbide. Observe that the pearlite nose shifts from 1 to more than 1000 s due to the effect of alloying elements in solid solution. Also notice that the martensite temperature increases after the precipitation of proeutectoid carbides in the high-alloy steel, as an effect of the decrease of alloying elements in solid solution. Curves were drawn from data in Reference 9 for the 1080 carbon steel and References 10 through 12 for 8%Cr steel (commercial names Sleipner, K340, and VF800AT, respectively).

As a background, the cooling curves shown in Figure 3.15 are also important to a discussion on the volume effect on the quenching technology and the consequences on *distortion*. When the cooling is accelerated, by using oil or water, the difference in temperature between the surface and core is increased. It is easy to show that the internal stresses developed by quenching are also directly related to this temperature difference [20]. Therefore, the increase in cooling rate in large dimensions will lead to higher stresses during cooling, which can cause extensive distortion of tools. And, in addition to distortion, cracks can also occur when those stresses are combined with transformation stresses (from austenite to martensite) and applied in a highly brittle microstructure, such as the medium- or high-carbon fresh martensite formed in tool steels after quenching. So, although water cooling is faster and enables the full quenching of tool steels without the need for high alloying elements, it is not applied extensively because of distortion or cracking. Oil quenching is very common for tool steel blocks or semifinished tools, but control in the furnace atmosphere is important in the last case, to avoid decarburization (loss or carbon and strength at the surface) or oxidation. For this reason, and also for safety and environmental issues related to oil quench, vacuum austenitizing followed by high-pressure quenching is one of the most common heat treatments today for finished tools. In the past, salt baths were used quite extensively, but environmental pressures have made this kind of treatment much less common now.

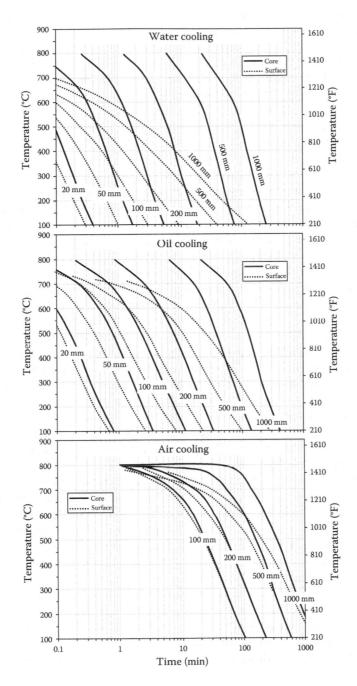

FIGURE 3.15 Variation of cooling time during water-, oil-, or air-quenching. Data show surface and core temperature profiles for bars of different dimensions. (From Bandel, G. and Haumer, H.C., *Stahl. Eisen.*, 84(15), 932, 1946.)

Before moving to the next topic, it is important to throw some more light on distortion, though this topic can also be found extensively elsewhere [20]. While distortion tends to increase with faster cooling, it is primarily determined by inhomogeneous cooling of the tool.* So, since fast cooling may be necessary and in most cases cannot be avoided, the distortion can be minimized by enabling a more homogeneous cooling, which can be done by better distribution of the coolant fluid (typically gas or oil) or by designing aspects to avoid that some segments of a given tool cool much faster than other segments. In oil quenching, the control of oil condition as well as the oil circulation inside the tank is important. And in vacuum heat treatments, that is, quenching by high-pressure nitrogen, the number and homogeneity of fans present inside the furnace are of high importance, in addition to the pressure applied.

Once the dimensions of a tool, the chemical composition of the tool steel, and its hardenability as a consequence are known, the quenching method and the cooling speed can be selected, using high-pressure gas, oil, or even air in specific cases. In terms of phase transformation, the initial driver is to avoid the formation of low-hardness constituents, such as pearlite for C above 0.3% or ferrite for lower carbon contents. While this is important to achieve the strength and is the primary point for low-alloy steels, it is not sufficient to promote adequate toughness, especially in tool steels alloyed with V, Mo, or W, via a phenomenon known as *quench embrittlement*. As shown in Figure 3.14, there is a dotted line in the continuous cooling transformation (CCT) diagram of tool steels, which represents the precipitation of carbides during cooling, but still in the austenite field. Another example for CCT diagrams of hot-work tool steels, for two different hardening temperatures, is shown in Figure 3.16. Alloy carbides (mainly V- or Mo-rich) tend to precipitate when they are found in a metastable condition, which occurs with the decrease in the austenitizing temperature down to typically 700°C. Below this temperature, the thermodynamic driving force still exists, but the kinetics are too slow to promote any substantial effect of quench embrittlement. This precipitation usually occurs at the austenite grain boundaries, because those regions are preferential sites for nucleation in terms of space and thermodynamic energy. As an overall effect, even in small amounts, the precipitation of such carbides may form semicontinuous (hard) distributions along the prior-austenite grain boundaries, which do not affect the hardness because of the small volume fraction but constitute an ideal path for crack propagation and therefore strongly reduce the final toughness. Figure 3.17 shows one example of several cases found in common failures of tools due to quench embrittlement, where toughness is reduced to one-fifth of the original value!

In Figure 3.16, one can observe an obvious dislocation of proeutectoid carbide bands when the austenitizing temperature increases. The reason is that an increase in austenitizing temperature leads to higher dissolution of carbides, and so the metastability of this microstructure will increase when the temperature is decreased. In other words, during quenching, the alloy elements will be more prone to combining

---

* Fast cooling by itself does not cause distortion, but an inhomogeneous cooling does. However, when fast cooling is applied, there is less available time to decrease the temperature differences along the tool, and so cooling becomes less homogeneous and distortion arises. That is the reason why it is common to see the correlation between fast cooling and distortion.

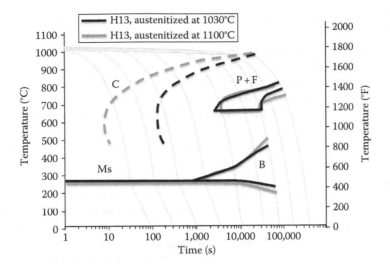

FIGURE 3.16 Superimposed CCT curves for H13 tool steel, austenitized at two different temperatures. Observe how the precipitation of proeutectoid carbides (C) advances to short times when high temperature is used. Other symbols: P, pearlite; F, ferrite; B, bainite; Ms, martensite. (Adapted from Ohlich, J. et al., *Atlas zur Wärmebehandlung der Stähle, Bd. 3, ZTA-Schaubilder*, Verlag Stahleisen mbH, Düsseldorf, Germany, 1973.)

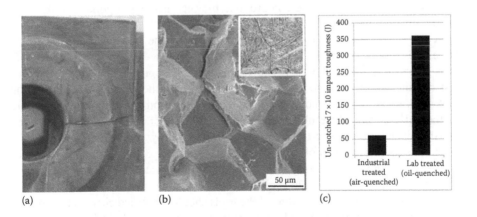

FIGURE 3.17 Example of premature failure in a modified H10 tool steel, in which the primary cause was attributed to quench embrittlement. (a) Picture, with ~5× magnification. (b) Example of the fracture analysis of the failure tool, with an insert of the optical microscopy image (same magnification as the fracture), showing, respectively, the intergranular fracture and the carbides in austenite grain boundary. (c) Change in impact strength after reheat treatment (both tempered to 45 HRC hardness, after quenching). (Adapted from Canale, L.C.F. et al., Failure analysis of heat treated steel components, in: *Failure Analysis in Tool Steels*, Mesquita, R.A. and Barbosa, C.A., eds., American Society for Metals (ASM), Materials Park, OH, 2008, vol. 1, Chapter 11, pp. 311–350.)

with carbon to form proeutectoid carbides, because their higher content in solid solution. Therefore, the precipitation of carbides on the grain boundary is pushed to shorter times and may easily occur even after fast cooling. The worst situation is when incorrect procedures of both austenitizing and quenching are combined, which intensifies dramatically in quenching embrittlement. This is the case in Figure 3.17, where the strong reduction of toughness was caused by a combination of a high austenitizing temperature (1100°C instead of 1020°C) and slow cooling (air-cooling instead of oil-cooling).

In addition to this strong effect on toughness, the concern about quench embrittlement is quite high today and can be explained by a set of factors. First, in most tools and dies, *quality control* is related only to the final achieved hardness. Therefore, the presence of quench embrittlement cannot be prevented in most cases and may be identified only after tool failure. A second point is the inverse correlation between cooling speed and both *quench embrittlement* and *distortion*. While distortion tends to increase with faster cooling (because the unavoidable inhomogeneities will show stronger differences under high cooling rates), the cooling rate must be fast enough to avoid or at least decrease the degree of quench embrittlement. For example, in the new recommendation of North American Die-Casting Association (NADCA) [4], the "Minimum quenching rate shall be: 50°F/minute (28°C/minute) between 1885°F and 1000°F (1030°C and 540°C), measured by T_s (in other words, T_s shall reach 1000°F (590°C) in less than 18 minutes)." In addition, NADCA also recommends that the temperature difference between the surface and the core must be controlled, to avoid distortion: "NOTE: An interrupted quench is recommended when the difference between the surface (T_s) and core thermocouple (T_c), exceeds 200°F (90°C)." This is a good example of the compromise between achieving good metallurgical quality and dimensional quality.

A third aspect related to quench embrittlement is the inevitable occurrence in large tools and therefore the changes required in *design*. As shown in Figure 3.15, even when large pieces are cooled in water, the core region is not able to achieve fast cooling because the cooling rate is controlled by the heat transfer inside the tool steel, regardless how cold the surface may be. In other words, the quenching method affects only the tool surface, and the cooling of interior regions will primarily depend on the heat transfer, which depends only slightly on the tool steel used but strongly on the tool dimension (Note in Figure 3.15 the small differences between the core areas of large blocks quenching in water and oil or even when compared to air). Therefore, to increase the mechanical properties achieved after heat treatment, the working regions of the tools should be placed close to the cooling areas, meaning that premachined tools are preferable to cooled blocks when quench embrittlement is considered. This machining approach was the standard when the machining technologies were not able to machine hardened and tempered tools, but today it is quite common to see extensive machining of tool steels up to 50 HRC hardness, or even slightly higher. There are several advantages in doing that, because the manufacturing cycles can be reduced and tools can be prepared with much fewer logistical issues and lower cost. However, the disadvantages are related to the quench embrittlement, because the core regions of a tool, with slow cooling and carbides on grain boundaries, become the working surfaces of the tools and so the performance can be compromised. There are cases where this effect is not strong, such as in low-alloy tool steels (with low V, Mo, or W), which are commonly machined from pre-hardened blocks. Examples of those grades are DIN 1.2738 or DIN 1.2714 (ASTM P20 modified or 6F3). However, high-alloy steels such as the 5%Cr tool steels (DIN 1.2343 or 1.2344/H11 or H13) are quite sensitive to this quench embrittlement and should be machined only from prehardened blocks when toughness is not important to the tool application or when the dimensions are small, with thickness less than 100 mm (4 in.).

The above-described aspects are most important for the quenching aspects and the final properties of tool steels. These aspects are the consideration of what hard-enability, tools size, and design have on the distortion, the final hardness, and the achieved toughness while considering the possibility of quench embrittlement. However, before moving forward, it is important to consider the microconstituents formed after quenching in terms of martensite or bainite and the presence of retained austenite. Also, because of low-alloy steels, it is a common belief in the industry that the final microstructure after proper quenching is 100% composed of martensite. While this is correct in principle, the large dimensions and the alloy contents exist-ing in tool steel promote considerable changes, and the final microstructure is more complex. For example, in low-alloyed tool steels, it is common that the final micro-structure is mostly composed of bainite, which achieves elevated hardness and also may promote precipitation hardening (secondary hardening, explained in Section 3.2.2.3) after tempering, a situation very similar to martensite. Large plastic mold steel or forging die blocks, made of DIN 1.2738 or DIN 1.2714, are typical exam-ples of applications where the microstructure is practically 100% bainitic. When observed under the microscope, bainite is usually coarser than martensite, although in medium- to high-carbon steels the differentiation between the two phases is often difficult microscopically. Although some sources in the literature discuss the pos-sible disadvantages of bainite over martensite, such as in 5%Cr hot work steels [22], this effect is not clear or proven. What is clear, especially in high-alloy steels, is that slow quenching is related to the decrease in toughness. And, while some authors may attribute this to the large amount of bainite, the low toughness is mostly caused by the presence of carbides in grain boundaries, or, in other words, by the quench embrittlement discussed earlier.

In high-alloy tool steels, such as high-speed steels or high-carbon cold work steels from D- or O-series, the final microstructure may also present retained austen-ite in addition to martensite or bainite (see Section 5.3.1). Retained austenite exists in those steels because the high carbon and high alloy content, which is important to increase hardenability, also promotes a decrease in the martensite transformation temperatures, bringing the finished martensite temperature below room tempera-ture. In practice, this means that part of the microstructure will not fully transform to martensite or bainite and will remain as austenite (then named *retained austenite*, see Figure 3.22). This phase is usually undesirable because it has low hardness and is unstable, causing variations in the final hardness of the tools as well as dimen-sional variations during the tool use. This last aspect is due to the fact that aus-tenite, being unstable at room temperature, tends to transform to martensite after stresses are applied, which is typically the working condition of the final tools. Once transformed to martensite, if in considerable volume, the retained austenite causes dimensional changes in tools, which is critical for applications such as fine blanking or punching tools. In addition, the martensite emerging from retained austenite is not tempered (often named *fresh martensite*), and the absence of tempering* makes it brittle, which may result in the premature failure of tools. As discussed in the next section, high-temperature tempering treatments, typically above ~500°C, may reduce the volume percentage of retained austenite when in reasonable amounts. However, an excessive amount of retained austenite cannot be reverted to martensite, which may result from incorrect or uncontrolled austenitizing conditions, especially in high-alloy and high-carbon tool steels. Under high austenitizing temperatures, the dissolution of primary carbides brings excessive amounts of alloy elements and carbon into solid solution, decreasing the M_s temperature and increasing the retained austenite, as shown in Figure 3.18 for D2 (DIN 1.2379) tool steel. A similar effect on retained austenite can be observed in Figure 3.13 in a T1 high-speed steel when the hardening temperature exceeds the usual value (Figure 3.13c and d). Therefore, for reducing the amount of retained austenite, the austenitizing conditions should be controlled, especially by using the temperatures recommended for each tool steel (Table 3.3).

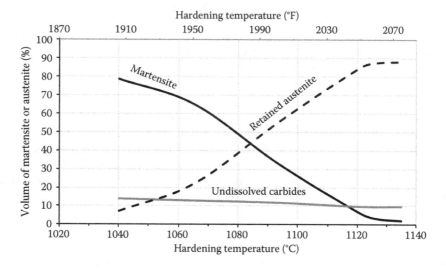

FIGURE 3.18   Changes in the volume fraction of microconstituents after quenching in a D2 tool steel after hardening from different austenitizing temperatures. Observe that the increase in austenitizing temperature leads to stronger dissolution of carbides, which increases the amount of C and Cr in solid solution, which in turn decreases the M_s and M_f temperatures and then increases the amount of retained austenite. (From Averbach, B.L. et al., The effect of plastic deformation on solid reactions, Part II: The effect of applied stress on the martensite reaction, *Cold Working of Metals*, American Society for Metals, 1949.)

---

* The effect of tempering treatment is discussed in Section 3.2.2.3; it usually leads to a decrease in hardness but a substantial increase in toughness.

So, in summary, the final microstructure for several large-dimension tools may present small or large amounts of bainite, in addition to martensite, which usually does not affect the strength or toughness. On the other hand, if pearlite or ferrite is formed, the strength is compromised, and a minimum cooling rate should be applied to avoid these constituents. If strong precipitation in grain boundary is present, toughness is also compromised (quench embrittlement) and, again, the quench rate should be increased or the design changed. Finally, mainly in high-carbon steels (C > 0.4%), the amount of retained austenite also should be controlled to avoid hardness or dimensional instability problems.

3.2.2.3 Tempering

After hardening, the tools are not yet ready for usage because they present a high level of internal stresses and very low toughness. Therefore, tool steels are hardly ever used in the hardened conditions and must be submitted to a further treatment, to adjust hardness and especially to increase toughness. This treatment is named *tempering*, and is typically conducted between 200°C and 650°C (390°F to 1200°F), depending on the desired final hardness and the working temperature of the final tool. Table 3.3 presents the hardness obtained after different tempering temperatures for the most common tool steels.

Tempering is a rather simple treatment, consisting of heating to the treatment temperature, usually in open furnaces,* followed by air-cooling. Although no complexity is found in the procedure itself, tempering is very important because it brings the last metallurgical transformation in tool steels, and mistakes in this step cannot be solved before tools are put to use. The mechanisms in tempering are also not trivial, and it is important to understand them so that the procedures can be properly set. Therefore, this section will be dedicated to specific tempering aspects, mainly those closely related to the final quality. It addresses the quality issues, in terms of (1) the decrease in hardness and the mechanism of secondary hardening, (2) the implications on hot work and high-speed steels, (3) changes in retained austenite, and (4) the need for multiple tempering treatments.

In short, tempering may be defined as a treatment to achieve, in hardened steels, decrease in hardness and an increase in ductility and toughness. While this is not incorrect, it does not show all the important specific aspects of tempering. It is not the purpose of this book to discuss in depth the reactions involved in tempering, but some aspects are extremely important to tool steels, such as the basics of carbon diffusion from martensite structure, the formation of cementite, and, in alloyed tool steels, the precipitation of secondary carbides.

It is important to first consider that martensite is a metastable structure, meaning that it is not the phase with lowest free energy, and thus with tendency to convert into new structure if sufficient time and temperature are allowed. So, if an as-quenched tool steel is subjected to higher temperatures for times ranging from minutes to hours, carbon will move away from martensite and form carbides. Martensite will then gradually change from the tetragonal structure to

---

* Without atmosphere control, although cases of partial or complete atmosphere control exist for tools needing surface protection.

cubic structure, that is, it will be converted to ferrite. In an overall view, carbon diffusion during tempering aims to convert the martensite structure into carbides and ferrite, which is the equilibrium structure in tool steels.

Since carbon in martensite solid solution is the main cause for *martensite hardness*,* hardness decreases when carbon moves away from martensite, which explains the decrease of hardness when tempering is applied. After diffusing away, carbon forms iron carbides (cementite or other transient iron-based carbides[†]) and those carbides increase in size and decrease in number with continued tempering; and consequently their ability to block dislocations decreases. The higher the temperature and the longer the time, the more extensive will be the diffusion of carbon and the coarsening of carbides. As a final result, martensite hardness continuously decreases by increasing the tempering time and temperature in carbon steels.

On the other hand, as shown in Table 3.3 and schematically in Figure 3.19a, alloyed tool steels present, in higher or smaller degree, an increase in hardness when submitted to tempering treatments in the range 400°C–600°C. As explained earlier, martensite hardness continuously decreases when the temperature increases, and this increase in hardness cannot be explained by martensite hardness. This second step of hardening during heat treatment (the first step being the martensite formation) is then called *secondary hardening*. Different from carbon, alloy elements presented in solid solution become "mobile" only at temperatures above 400°C[‡] and, when able to diffuse, they react with carbon, forming different types of carbides, called secondary carbides, because they are not formed directly from liquid but from solid-state precipitations. The increase in hardness with the precipitation of alloy carbides depends on their ability to block the movement of dislocations and then basically depends on the number of carbides and their size, which is usually in nanoscale.

At lower temperatures, carbides may not exist or may be present in very small amounts to block the movement of dislocations, so the hardness contribution is small. With the increase in temperature or time, carbides increase in volume fraction and in number, leading to a continuous increase in hardness until a maximum point is reached and the coarsening of carbides start to dominate. Coarsening is a natural phenomenon to decrease surface energy by converting two or more small particles into one bigger particle, and will occur with the increase in time or temperature. Coarsening then reduces the number of carbide particles, which in turn decreases the number of pinned dislocations by precipitates and finally decreases the strengthening. So a peak in hardness is reached when maximum precipitation exists and before appreciable softening effects take place by coarsening. Beyond

---

* Martensite is a structure with very high concentration of dislocation: about 10^{13} cm of dislocations per cubic centimeter (usually noted as $10^{13}/cm^2$), while an annealed steel has only $10^5/cm^2$, which is ~10–100 million times less. When carbon is in solid solution in martensite, it is usually located close to dislocations, stabilizing their positions and then delaying their movement. The net effect is that the strength of martensite is directly dependent on the carbon in solid solution.

[†] Before cementite, transient carbides are usually formed, such as epsilon (ε) carbide and eta (η).

[‡] Because alloy elements are substitutional elements, the diffusion within the microstructure is much slower and exists in practice only after temperatures of 400°C or more. Carbon is interstitial (see Reference 1 for definitions), and its diffusion is several orders of magnitude higher.

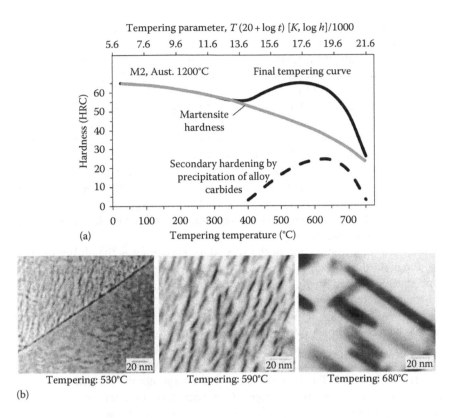

(a)

(b)

FIGURE 3.19 (a) Tempering curve for M2 high-speed steel, showing the contribution of martensite and secondary carbide precipitation to final tempering curve. The curve was obtained by merging the tempering curve of O1 tool steel (~0.9%C) as a martensite matrix only, and the final curve is the actual M2 tempering curve; the precipitation curve was obtained by subtraction: the final martensite curve (Based on Böhler Edelstahl, *Technical Catalogue for Tool Steels*, Available online at: http://www.bohler-edelstahl.com/english/1853_ENG_HTML.php, Accessed December 2015; Villares Metals, Technical catalogue for tool steels, Available online at: http://www.villaresmetals.com.br/pt/Produtos/Acos-Ferramenta, Accessed December 2015), from tempering curves of M2 high-speed steel and O1 tool steel to represent the unalloyed martensite hardness. (b) Examples of secondary carbides for different tempering temperatures. (From Ebner, R. et al., Methods of property oriented tool steel design, *Proceedings of the Fifth International Tooling Conference*, Leoben, Austria, 1999, pp. 3–24.)

the peak, carbides start to be coarser, and the hardness continuously decreases.* Figure 3.19b displays these changes in carbides for a high-speed steel tempered for different times and temperatures. Both time and temperature may then be correlated to the Hollomon–Jaffe parameter [25] $P = T(20 + \log t)$, with T being the temperature

---

* If very long tempering at high temperatures is applied, the coarsening reaches a point where carbides are visible in optical microscopes, in overtempered condition, and hardness may be really small. This is actually the subcritical or temper annealing treatment, described in Section 3.2.1.

in kelvin and t the time in hours (check second x-axis on Figure 3.19a). Because of the logarithmic effect, it is clear that the effects of temperature are much stronger than those of treatment time, and it is common to see tempering charts, like the one shown in Figure 3.19a, with only hardness versus tempering temperature, for a fixed time, usually of 4 h (2×2 h).

The fine alloy carbides that precipitate during tempering are more effective when they are of size of a few nanometers, which are usually observed in V-, Mo-, and W-alloyed tool steels. Cr also forms secondary carbides, but their contribution to secondary hardening is smaller than that of V, Mo, and W carbides at high temperatures (above 550°C), because they precipitate at lower temperatures. For reference, Figure 3.20 shows the separate effect of all these alloy elements on the secondary hardening of a medium-carbon steel (~0.50%C). This explains why tool steels that are very dependent on secondary hardening usually contain high levels of Mo and V, or W in some alternatives.

One important implication of the secondary hardening is the ability of steel to retain high hardness after exposure to high temperatures, which is essential for hot work and high-speed steels often exposed to >500°C during use. This ability to withstand high temperatures without losing strength is frequently referred to as *tempering resistance*, and the high alloy content in several hot work and high-speed steels is added to improve it. In fact, it is not an exaggeration to say that without secondary hardening tool steels cannot be used at high temperatures. One interesting example of this dependence is the history of the development of high-Cr, high-C tool steels (D-series): These steel types form a large number of primary carbides and are very effective for room-temperature wear resistance, and were developed to substitute high-speed steels in cutting tools at a lower cost due to the lower amounts of Mo, V, and W compared to M2. However, these last three alloying elements are essential for secondary hardening, and their low contents in D-series tool steels make these compositions usable only at room temperature and they are today the basis of cold work steels.

Therefore, overall, tempering is composed by a series of solid-state reactions in which the microstructure is converted from a metastable martensite structure, with carbon and alloy elements in solid solution, to a more stable microstructure, composed of low-tetragonality martensite and, with further increase in temperature, to ferrite. Iron carbides are initially formed, and at higher temperatures (typically above 400°C), alloy carbides precipitate in the microstructure. Figure 3.21 presents a summary of the hardness development during tempering and the respective carbides that precipitate during tempering of H11 tool. Observe the continuous decrease in martensite hardness, which is combined with the hardness peak promoted by the alloy carbides and then leads to a hardness peak or plateau until 550°C. The final effect is that the precipitates avoid a fast decrease in hardening, which would occur if only martensite was present and then high tempering resistance of this hot work steel.

One further important phenomenon also occurs in tool steels tempering: The *decomposition of retained austenite* and, because of this, the need for *multiple tempering* treatments. As explained earlier, austenite starts to convert to martensite when the M_s temperature is reached, since this transformation is dependent not on diffusion but only on the achieved temperature. To convert all the microstructure

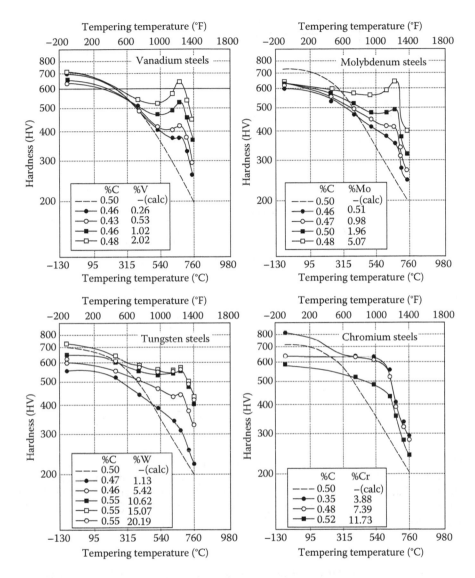

FIGURE 3.20 Effect of alloying elements on the secondary hardening during tempering, leading to an increase in hardness due to precipitation of carbides. (Adapted from Crafts, W. and Lamont, J.L., *Trans. TMS-AIME*, 180, 471, 1949.)

to martensite, the final martensite temperature (M_f) should be reached, which does not occur because M_f is below room temperature for most high-alloy steels (due to the high amount of carbon and alloy elements in solid solution). However, the retained austenite may transform after tempering, because the precipitation of carbides during tempering (shown in the last paragraph and summarized in Figure 3.21) decreases the amount of carbon and alloy elements in solid solution, thus increasing M_f temperatures. After cooling down, part or all of the retained austenite may

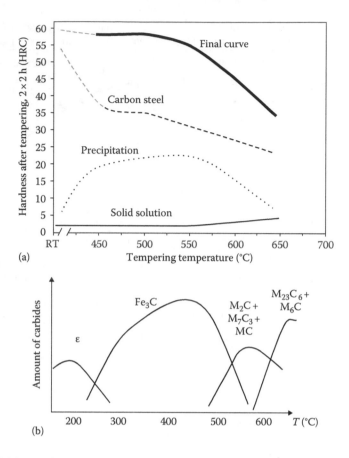

FIGURE 3.21 (a) Quantitative calculation of the martensite, solid solution, and precipitation hardening effects on the final hardness of H11 tool steel. (Adapted from Mesquita, R.A. and Kestenbach, H.J., *Mater. Sci. Eng. A*, 566, 970, 2012.) (b) Sequence of precipitation of a low-Si H11 tool steel showing the types of precipitated carbides. (From Banerjee, B.R., *J. Iron Steel Inst.*, 203, 166, February 1965.)

then transform to martensite once the M_s and M_f temperatures have increased by the decrease of alloying elements in solid solution. Therefore, after the first tempering, the microstructure in high-alloy tool steels is often composed of not only tempered martensite but also untempered martensite, also known as *fresh martensite.* The transformation of retained austenite to martensite during tempering is actually desired and is a very effective way to reduce or eliminate the negative effects of retained austenite described earlier (dimensional instability and brittleness in use). However, the microstructure after only one tempering will contain amounts of fresh (brittle) martensite within the tempered (and tougher) martensite, and even a small amount of fresh martensite may compromise the toughness of the overall tool steel. Therefore, a second tempering treatment is recommended to eliminate fresh martensite and bring the final state to a fully tempered martensite microstructure.

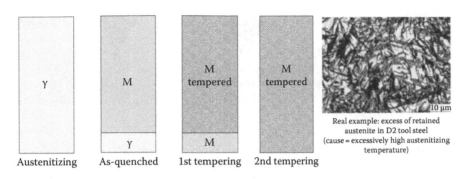

Real example: excess of retained
austenite in D2 tool steel
(cause = excessively high austenitizing
temperature)

| Austenitizing | As-quenched | 1st tempering | 2nd tempering |

FIGURE 3.22 Scheme of the austenite, martensite, and tempered martensite as a function of the number of tempering treatments. In reality, the retained austenite, or the fresh martensite formed from it, is distributed within the needles/laths of tempered martensite, as shown in the micrograph (light matrix is mostly retained austenite).

In cases of high-alloyed high-speed steels, a third tempering may be necessary to eliminate traces of fresh martensite after the second tempering.

Figure 3.22 shows a scheme of all those changes promoted by retained austenite and the need for multiple tempering treatments. It is always recommended that both tempering treatments are applied at similar temperatures. Typically, if the hardness is already achieved after the first tempering, the second tempering is done at ~50°C (90°F), 90°F lower than the first tempering. Alternatively, if the hardness is still high after the first tempering, the second tempering is used to adjust the hardness. This option is actually preferable, since it will lead to microstructures that are more alike.

3.3 UNDISSOLVED CARBIDES AND THEIR EFFECTS ON TOOL STEEL PROPERTIES

3.3.1 THE IMPORTANCE OF UNDISSOLVED CARBIDES IN TOOL STEEL MICROSTRUCTURES

For most applications in engineering, the design of a proper microstructural matrix is sufficient for achieving different combinations of strength and toughness. The same is also valid for some tool steels, for example, those used in plastic molding or hot working tools. However, in conditions of high wear exposure, the strength of a microstructural matrix—tempered martensite in the case of tool steels—is not sufficient to withstand the abrasion and promote adequate die life. This occurs because most abrasive agents causing tool wear are nonmetallic, typically ceramic particles. Examples of typical pure abrasive conditions are those existing in tools used to shape ceramic bricks, such as refractories, or mining. Figure 3.23 shows a schematic example of an abrasive particle sliding against a tool. Because of their ceramic nature, those particles typically have hardness of 1500 HV or more and penetrate into the tool steel (usually with hardness between 500 and 900 HV). Therefore, as shown in Figure 3.23, particles will easily penetrate the tool surface and will move and wear out small amounts of material at the tool surface during tooling operations.

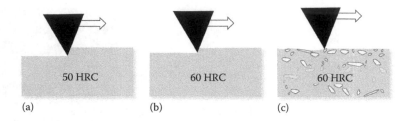

FIGURE 3.23 Effect of hardness and the addition of carbide particles on the abrasion of a tool steel surface. The increase in hardness, from (a) to (b), leads to less penetration of the particles and reduces the wear. But the best abrasive resistance is obtained when undissolved carbides are distributed in a high hardness matrix, shown in (c). (Modified from Canale, L.C.F. et al., Failure analysis of heat treated steel components, in: *Failure Analysis in Tool Steels*, Mesquita, R.A. and Barbosa, C.A., eds., American Society for Metals (ASM), Materials Park, OH, 2008, vol. 1, Chapter 11, pp. 311–350.)

After several hours or days of operation, this wearing mechanism will cause permanent change in the shape of the tool, which leads to either improper geometries of the produced parts or loss in productivity. Increasing the tool hardness (as shown in Figure 3.23a to b) will increase the resistance to penetration of particles, and as a final result, the amount of material lost by wear will decrease. However, there is a limit for the hardness of tool steels: For example, the maximum hardness of martensite, even after strong martensite and precipitation hardening, is limited to ~950 HV. In this case, the abrasive wear cannot be stopped by only increasing the tool hardness.

A third option for increasing the abrasion resistance is shown in Figure 3.23c: namely "adding inside the tool steel" particles of a ceramic nature, which can be as hard as or even harder than the abrasive particles (see Figure 3.24). Those added particles are actually carbides that are not dissolved during hardening; in fact, they are designed not to dissolve and to be present within the martensitic matrix in sizes of 1 μm or larger. The existence or absence of undissolved carbides depends on the type of carbides, determined by the correspondent alloying element, the amount of carbon and alloy content, and the hardening temperature, as described in detail in Section 3.2.2.1 and Figure 3.11. It was shown in Figure 3.11 that the alloying elements show very different limits of solubility because of the chemical attraction between those elements and carbon, which is actually represented by the free energy of formation (shown in Figure 3.10). For example, the carbides that are more easily dissolved are the Cr carbides, followed by the Mo carbides. Carbides that are more resistant to dissolution are the MC-type carbides; the Nb-rich MC carbides more resistant to dissolution than the V-rich MC carbides.

The carbides mentioned here are usually of the same nature and crystal structure as the small precipitates shown in Section 3.2.2.3, but their function in the microstructure is completely different. The small precipitates formed during tempering have dimensions in the nanoscale and are intended to pin and delay the movement of dislocations, thus increasing the strength. The particles shown in this section, on the other hand, have sizes of 1–100 μm (easily observable under an optical microscope)

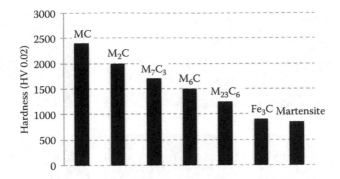

FIGURE 3.24 Typical hardness of different types of complex carbides. The carbides are shown in the M_xC_y form, where M is usually a mixture of Fe and alloying elements and never a pure carbide. However, each stoichiometry is based on the crystal structure and usually has a high content of one of the alloying elements: $MC \rightarrow Nb$ or V; M_2C and $M_6C \rightarrow W$ or Mo; $M_{23}C_6$ and $M_7C_3 \rightarrow Cr$; $M_3C \rightarrow Fe$. (From Brandis, H. et al., Metallurgical aspects of carbides in high speed steels, *Processing and Properties of High Speed Tool Steels*, Wells, M.G.H. and Lherbier, L.W., eds., TMS-AIME, Warrendale, PA, 1980, pp. 1–18; Tarasov, L.P., *Metal Prog.*, 54(6), 846, 1948; Elsen, E. et al., *Metall*, 19, 334, 1965; Brook, G.B. and Crompton, J.M.G., Fulmer Report R 319/4, Fulmer Research Institute, 1971, Apud: Karagoz, S. and Fischmeister, H., *Metallurg. Trans. A*, 19Z, 1935, 1988.)

and have the role of avoiding the penetration and, consequently, reducing scratching damage on the tool surface. High microstructural hardness is still important because a hard matrix is necessary to hold the undissolved carbides, which will then play a role in avoiding the material removal by abrasive agents and, as a final consequence, increase the wear resistance. Therefore, practically all tools used in highly abrasive conditions, such as high-speed and cold work tool steels, have a large amount of particles, usually up to a maximum of 20 vol.% (but usually less than 15 vol.%). The addition of more particles is in theory possible, just by adding carbon and alloying element, but the microstructure will reach such a high level of brittleness that the steel cannot be processed in a steel mill by hot-forging or rolling. In addition, if a very high volume fraction of particles is present, the tool steel may not have sufficient toughness to avoid cracking during tooling. However, within the normal amounts, that is, up to 15 vol.%, the undissolved carbide particles are an extraordinary mechanism to increase the wear resistance of tool steels, especially under abrasive conditions.

3.3.2 BALANCE BETWEEN WEAR RESISTANCE AND TOUGHNESS

3.3.2.1 Wear Resistance

As mentioned earlier, the presence of particles and the increase of wear resistance are usually desired for cold work and high-speed tool steels. Before describing the details of the effects of undissolved carbides in wear resistance, a general overview on the microstructure of representatives of those steels is given in Figure 3.25. The first thing to realize is that all particles are bright in the microstructure. This means

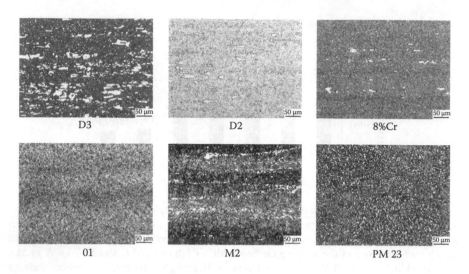

FIGURE 3.25 Microstructure of typical cold work and high-speed steels, all in bars with ~60–90 mm diameter. (From Mesquita, R.A. and Barbosa, C.A., *Technol. Metal. Mater.*, 2, 12, 2005.)

that they were not etched by the acid during sample preparation for optical micros-copy, which in turn is a result of their ceramic nature (only special etchings, i.e., spe-cial type of acid corrosion, are able to affect the carbides). The matrixes, however, vary from one type of steel to another as a result of different heat treatments. More important than the etching conditions, it is easily observed that the microstructures differ substantially in terms of size, amount, and distribution of particles. For exam-ple, in relation to the cold work steels, the O1 tool steel shows very small amounts of carbides, which increase to 8%Cr, followed by D2 and D3 steels. Two high-speed steels are also shown in Figure 3.25—M2 and PM M3:2—the first produced by con-ventional casting and the second by powder metallurgy. As described in Chapter 2 on manufacturing, the PM grade is obtained with a very fast solidification method, which converts the melt almost instantaneously to small powder particles, and then the structure is refined, leading to a much more homogeneous microstructure.

The differences in the number and distribution of particles will determine why and where those steels are typically used. As explained in detail in the following paragraph, the increase in the number of particles leads to better wear resistance, but considerably reduces the toughness of the tools. In some situations, a tool steel with higher wear resistance cannot be used because toughness is impaired and micro or macro cracking may occur. In the case of O1 tool steel, the microstructure is practi-cally only hardened martensite and is solely responsible for the wear resistance. The addition of high Cr and high C contents, from the 8%Cr steels to D2 steel and finally to D3 steel, leads to an increase in the number of particles, which increases the wear resistance but also decreases the toughness. For example, extremely high abrasion resistant tools are often produced with D3 tool steel, but in several metal-forming operations, where toughness is also important, tools made of D2 or 8%Cr steels are

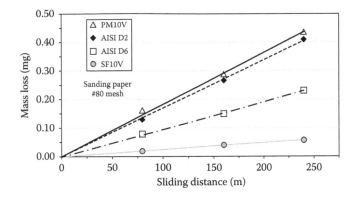

FIGURE 3.26 Pin-on-sandpaper test to evaluate the wear resistance of different tool steels, all hardened and tempered to the same hardness of ~60 HRC. Observe that the higher the mass loss, the lower the wear resistance; so, in this case, the wear resistance increased in the following order: PM10V, D2, D6, and SF10V. SF10V and PM10V have the same chemical composition but were produced by different casting techniques: spray-forming and powder metallurgy, thus leading to different sizes and distributions of carbides. D2 and D6 were conventionally cast and hot-rolled. (From Mesquita, R.A. and Barbosa, C.A., *Proc. Euro PM 2004 (Powder Metall. World Cong. Exhibition)*, 5, 53, 2004.)

more common today. Lastly, by also having very high wear resistance, M2 steel is very common in cutting and forming tools with higher hardness (usually 65 HRC vs 60 HRC) and harder undissolved carbides (MC and M_6C versus M_7C_3) compared to D-series steels.

The addition of particles leads to an obvious increase in wear resistance, both abrasive and adhesive wear. Starting with abrasive wear, Figure 3.26 shows the comparison of three different grades, on the wear loss of the hardened tool steel against a #80 mesh sandpaper. D2 and D6 (similar to D3) microstructures are shown in Figure 3.25, and it is seen that the amount of undissolved carbides in D2 is ~13 vol.%, while the volume fraction of carbides in D6 is ~20 vol.%. This explains the obvious increase in wear resistance going from D2 to D6. However, the steel SF10V, with ~15 vol.% carbides, is far better than D6 in terms of abrasive wear resistance. The microstructure of this steel is shown in Figure 3.27a for different regions of the ingot. Despite not being a commercial product, this steel represents a very good example for learning purpose and is discussed here. The first factor of the higher wear resistance is the presence of a large volume of carbides of MC type, which are almost twice as hard as the M_7C_3 carbides found in the D-series steel (see Figure 3.24). So wear resistance increases by the increase in carbide hardness. A second important effect on SF10V is the distribution of carbides, which is very homogeneous as a result of its production method.*

---

* SF means spray-forming, which is an alternative method researched in the literature for several alloys. It was used industrially for tool steels only for a few years, but the commercial availability today seems to be reduced. Basically, it consists of spraying the liquid steel, in this sense similar to powder metallurgy, but capturing the droplets on a substrate before they become solid. The solidification speed is intermediate between those of PM and conventional casting, leading to larger carbides than PM but with better distribution than conventional. Check References 34 and 35.

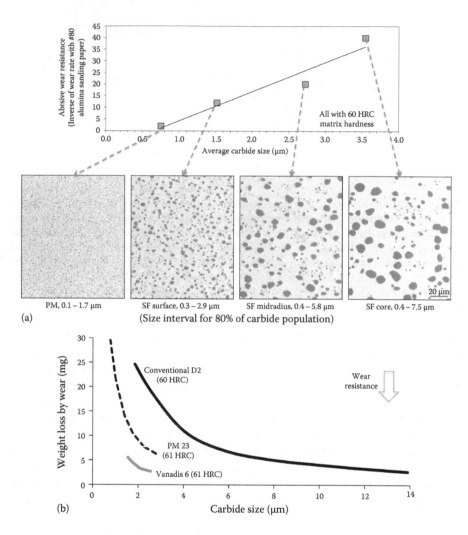

FIGURE 3.27 Examples of carbide size and abrasive wear resistance. (a) 10%V cold work tool steel grade (2.6%C, 5%Cr, 1.4%Mo, 10%V) produced by different casting methods (PM = powder metallurgy or SF = spray-forming) and samples taken from different regions referred to the original ingot, leading to different carbide distributions. Observe that coarser carbides, when well distributed, lead to better abrasive wear resistance (Modified From Mesquita, R.A. and Barbosa, C.A., *Proc. Euro PM 2004 (Powder Metall. World Cong. Exhibition)*, 5, 53, 2004), with further content added by the present author on the surface and core of SF 10%V cold work steel. (b) Variation of weight loss by wear (inversely related to wear resistance) as a function of carbide size for three different steels. Note that, as the carbides become coarser, the abrasive wear resistance increases. (From Grinder, O., PM HSS and tool steels—Present state of the art and development trends, *Proceedings of the Fifth International Conference on Tooling*, Loeben, Austria, 1999, pp. 39–47.)

Some scratched and partially polished microstructures are shown later in Chapter 5 (see Figure 5.6) showing that wear groves form preferentially on the areas free from carbides and therefore can remove more volume in the conventional steel.

A third important factor for abrasive wear resistance is that the carbides have to be reasonably large, similar to abrasive media. This can be anticipated in Figure 3.26, where SF10V and PM10V are compared. Both steels have exactly the same chemical composition and thus the same carbide types and volume fraction, but different manufacturing methods, with the carbides in the PM being 2–10 times finer. This fact is very clear in Figure 3.27a, which shows the PM10V and different regions of the SF10V, with different particle sizes: When the average size of the carbides increases, the abrasion resistance (against very coarse sandpaper) also increases. The same is shown for other PM high-speed steels in Figure 3.27b. In the last case, for the same size, the curves are also shifted as a result of the different chemical composition and consequently different types and hardness of the formed carbides.

In short, Figures 3.26 and 3.27 lead to the conclusion that the increase in abrasive wear resistance depends on, and increases with, three main factors: (1) the amount of undissolved carbides, (2) the size of the carbides, and (3) the homogeneity of the carbides distribution. The first two factors primarily depend on the chemical composition, that is, the amount of carbon and alloy elements in a tool steel. And the third factor depends on the manufacturing method, that is, the increase in homogeneity related to an increase in solidification speed (more precisely the cooling rate on solidification) and the amount of hot deformation applied to the tool steel during forging and rolling. These last concepts will be explained in Section 3.3.2.

It will be shown later, in Chapter 5, that another type of wear, called *adhesive wear*, occurs during metal–metal contacts. In some literature, this mechanism is also called galling or cold welding. This condition of wear is generated when high pressure is applied to two metals, which then tend to adhere in the contact areas. After the tool is opened and the part extracted, small pieces of the tool (that were adhering to the part) may be removed, especially very hard tools with low toughness. So the presence of carbides within the tool microstructure, by presetting a nonmetallic nature, reduces the contact area prone to adhesion and may also reduce adhesive wear during cold-forming (cold work steels) or high-temperature cutting operations (high-speed steels). However, the amount of particles and their distribution should be controlled: Concentrated particles will not act so well in reducing the metal–metal contact area once several areas of the tool steel contain no particles, and will also create a path for cracks propagation, facilitating the adhesive wear.

To summarize, abrasive wear depends on hard carbides, usually of large size (comparable to the abrasive agent), in large amounts and of high hardness. Adhesive wear resistance also depends on the quantity of hard carbides, but small and very well distributed carbides are desirable.

3.3.2.2 Toughness

Any kind of hard particles in a metal, of nonmetallic nature, leads to a decrease in the ability to withstand the applied forces before cracking. Following the traditional fracture mechanics theory, nonmetallic particles act like crack initiation sites, facilitating the fracture and thereby decreasing the toughness. In fact, the production of high-quality steels consists of applying melt-shop processes able to remove the maximum number of nonmetallic inclusions, such as slag and other oxide particles, and the hot-forming operations (typically by rolling) to remove any traces of porosities or internal voids. In cold work or high-speed steels, as explained earlier, nonmetallic hard particles are added on purpose, to increase wear characteristics. But, in the perspective of fracture mechanics, they continue to act as crack initiation sites, and there are studies showing that carbides and slag inclusions play similar effects in crack initiation and propagation. So, the addition of undissolved carbide particles to tool steels also decreases the toughness.

The explanation above is well exemplified by Figure 3.28a, where the same steel (and so the same type and amount of carbides) is produced with different casting

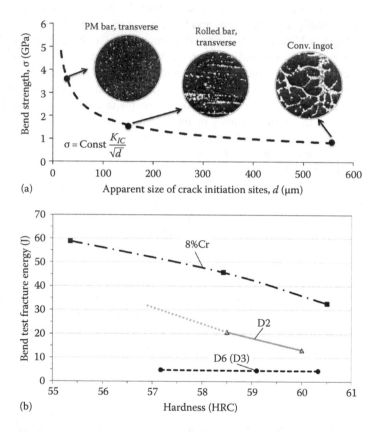

FIGURE 3.28 Examples of carbide size and distribution affecting the toughness of tool steels containing undissolved carbides, measured by the bend test. (a) Carbides that act like initiation sites for cracks and decrease the bend strength (and consequently the toughness) of high-speed steels. (Plot with data from Grinder, O., PM HSS and tool steels—Present state of the art and development trends, *Proceedings of the Fifth International Conference on Tooling*, Loeben, Austria, 1999, pp. 39–47.) (b) Energy absorbed for fracture during the bend test for different cold work tool steels and with different hardness. (Modified from Mesquita, R.A. and Barbosa, C.A., *Technol. Metal. Mater.*, 2, 12, 2005.)

and forming methods and tested in terms of fracture toughness. It is clear that an increase in the size of carbides leads to a strong decrease in the toughness, in this case measured by the bend test.* Comparing different grades, Figure 3.28b again shows steels D2, D3, and 8%Cr, for different hardness values, pointing that toughness increases from D3 to D2 and from D3 to 8%Cr, mainly as a result of the decrease in the amount of particles. However, in terms of 8%Cr, this is a cold work steel tempered at high temperature (with precipitation hardening), which leads to a reduction in martensite brittleness, and also contributes positively to increase toughness; in short, 8%Cr is tougher because of lower amount of particles and also because of a tougher microstructural matrix.

3.3.3 FORMATION AND FACTORS AFFECTING THE DISTRIBUTION OF UNDISSOLVED CARBIDES

3.3.3.1 Equilibrium Diagrams and Solidification

After knowing the importance of carbides for wear resistance and the drawback in terms of decreasing toughness, it is now important to understand how the carbides are formed in tool steels. In terms of chemical nature and properties, the particles in the microstructure of cold work and high-speed tool steels may be understood as ceramic reinforcements within a tougher metallic matrix, that is, typically a metal matrix–ceramic composite. While this would be correct in terms of mechanical behavior, the term "composite" is not commonly used to tool steels or other ferrous materials that present carbide particles, because the final metallic product is obtained not by the physical mixture and consolidation of metals plus ceramic particles but by the formation of the hard particles directly from the molten steel, during solidification.†

In Figure 3.29 pseudo-binary Fe–C diagrams are shown for M2 composition and for 13%Cr steels, the latter being the basis for D2 or D3 tool steels. Considering Figure 3.29a, the solidification of a typical M2 composition, with about 1%C, is highlighted by line A–B and is discussed in detail, as follows: Starting from the liquid, the first phase to be formed is ferrite (α), with very low carbon content (~0.05%C). As the α-phase grows in a dendrite morphology, the excess carbon and alloy elements are rejected from this phase and becomes concentrated in the liquid within the dendrite arms; although not precisely predicted by the diagram, one can note that the *pseudo-liquidus* line continuously moves to the right, pointing that the amount of carbon in the liquid is increasing. After crossing about 1350°C (2460°F), part of the α-phase also transforms to austenite (γ). Although austenite can dissolve more carbon than ferrite, the solubility of carbon is still limited and the liquid continues to be enriched in carbon and alloying elements. Just below ~1280°C (2340°F), the

---

* Hardened high-speed steel shows very low ductility, and in the bend test, the area under the stress–strain curve is proportional to the strength. So, the higher the bend strength, the higher the absorbed energy to breakage. More details on the bend test are given in Reference 38.
† Scientifically, the precipitation of carbides form the melt is often named "*in situ* metal matrix composite," but in industrial literature, the term "metal composite" is reserved only to a mixture of preexisting carbides or other ceramic particles to a metal.

composition is brought to the field $(\alpha+\gamma+C+L)$,* meaning that carbides start to form. Some carbides may precipitate in an isolated manner and directly from the melt, usually named *primary carbides*. Primary carbides are usually from the more stable carbide-forming elements, notably V or Nb. Once they grow freely in the liquid, they can assume the lowest surface energy condition and thus with a shape that tends to be with sharp facets or cuboid-type. The primary carbides are not the major part of carbides, and a fair amount of liquid with high carbon and high amount of alloying elements is still present after primary carbides precipitate. At about 1240°C (2270°F), the microstructure changes to a field with only $(\gamma+C)$, meaning that all the liquid is no longer present; this is obtained by the eutectic decomposition of the liquid, which has a very high concentration of carbon and alloying elements, forming lamellas of carbides and austenite. The carbides formed in this stage are then usually named *eutectic carbides*.

The same solidification sequence can be observed for 13%Cr tool steels, as shown in Figure 3.29b. One important difference may be noted: because of the carbon content, 1.5%C for D2 and 2%C for D3, the amount of primary carbides (formed in an isolated manner and directly from the liquid) is usually very small. Large amounts of primary M_7C_3 is observable only in compositions above 3.5%C, where the first phase

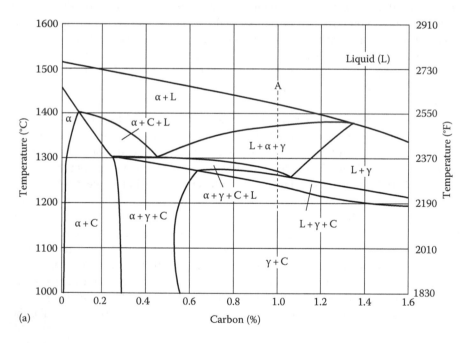

(a)

FIGURE 3.29 Pseudo-binary Fe–C diagrams, for compositions close to (a) M2 high-speed steel. (Adapted from Horn, E. and Brandis, H., *DEW Tech. Ber.*, 11, 147, 1971.) (*Continued*)

---

* The field $\alpha+\gamma+C+L$ is followed by $\gamma+C+L$ when all ferrite transforms to austenite, but this does not affect the carbide formation substantially.

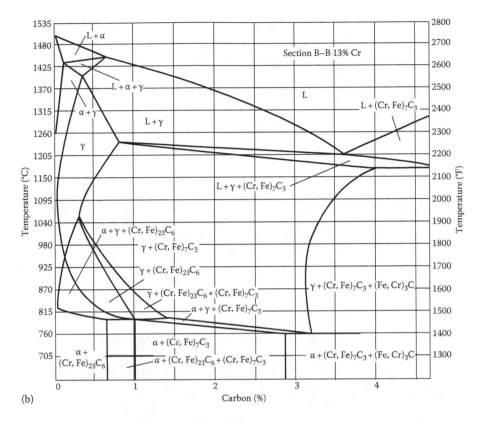

(b)

FIGURE 3.29 (Continued) Pseudo-binary Fe–C diagrams, for compositions close to (b) D-series cold work tool steels. (Adapted from Bungardt, K. et al., *Arch. Eisenhuttenwes.*, 29, 193, 1958.)

to form is the M_7C_3, instead of austenite. Those compositions, in the Fe–C–Cr diagram, are named *hypereutectic alloys*, and are not in the range of tool steels, because they are too brittle to be hot-rolled, and may be used as cast parts: *high-Cr white cast irons*. Hypoeutectic compositions in the Fe–Cr–C are, however, very common in D-series tool steels, and the mechanism is the same: once the solidification starts, the first metallic phase—austenite in this case—cannot dissolve all the carbon, and this element is then rejected to the solidification front, in a similar way as explained for high-speed steels in the last paragraph. When the temperature is low enough, the enriched liquid undergoes a eutectic reaction, converting all the liquid to a mixed structure of carbides and austenite, usually with lamellar morphology. The formed carbides are then also named *eutectic carbides*, rich in Cr and with M_7C_3 structure. Figure 3.30 shows in a schematic way the presence of the metal dendrites (MDs) and the distribution of carbides.

In terms of microstructure distribution of the formed carbides in high-speed steels, a typical example of such an as-cast microstructure of an M3 (type 2) steel is shown in Figure 3.31 (more pictures in Figure 3.33). The majority of eutectic morphology

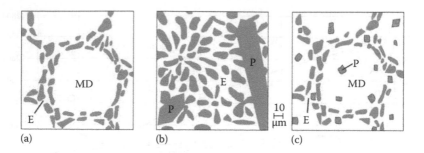

(a)   (b)   (c)

FIGURE 3.30 Schematic microstructure of wear-resistant white cast irons, showing different combinations of microconstituents: metal dendrites (MDs), net-like eutectic carbides (E), and hypereutectic primary carbides (P). The three microstructures show three composition ranges of high-chromium white cast irons. (a) hypoeutectic, carbon usually between 2% and 4%, with metal dendrites surrounded by eutectic carbides; (b) hypereutectic, usually with more than 4%C, with primary Cr-rich carbides formed previous to the eutectic carbides; and (c) again hypoeutectic composition, but with primary MC carbides formed prior to the metal dendrites and the eutectic carbides. (All from Berns, H., *Wear*, 254, 47, 2003.)

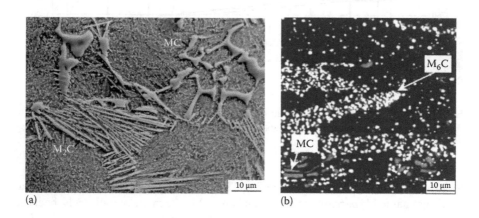

(a)   (b)

FIGURE 3.31 As-cast microstructure of M3:2 high-speed steels, showing (a) the as-cast microstructure, with eutectics of M_2C and MC carbides. Topographic image with scanning electron microscopy. (From Mesquita, R.A. and Barbosa, C.A., Hard alloys with lean composition, Patent: US 8,168,009, Priority, 2006.) (b) M_2C carbides decomposed into M_2C (bright phases) and MC (gray phases) carbides, compositional image from scanning electron microscope. (From Mesquita, R.A. and Barbosa, C.A., *Mater. Sci. Eng. A*, 384, 87, 2004.)

is composed of M_2C carbides, which during heating to hot rolling is converted to $MC + M_6C$. In Figure 3.31a, the eutectic carbides of MC and M_2C types are shown. In Figure 3.31b, the M_2C is shown broken to small particles of MC and M_6C, and the MC eutectics are also shown.

The undissolved MC carbides are present in high speed steels usually in smaller amounts than the M_6C carbides (see Figure 3.12 shown several sections before), depending mainly on the C, V, or Nb contents. For high V, one way to quantify the amount of primary carbides is to observe the solidification interval, which is defined as the range of temperature between the start of primary MC formation and the general conversion of the microstructure to eutectics. Observe in Figure 3.32a that the MC solidification interval increases continuously with the amount of V, starting from zero at about 1%V. This means that a high-speed steel with less than 1%V would have MC-type carbides only coming from the decomposition of M_2C eutectic or even from MC eutectics, but would not have

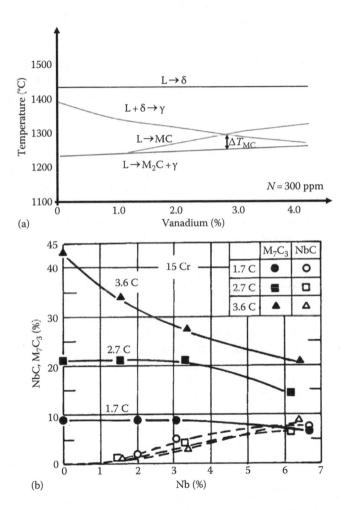

(a)

(b)

FIGURE 3.32 Formation of MC carbides in high-C, high-alloy steels. (a) Influence of V content on the precipitation of primary MC and the solidification interval for this carbide (ΔT_{MC}). (Plotted with data from Mizuno, H. et al., *Denki Seiko*, 55(4), 225, November 1984.) (b) Increase in the volume fraction of Nb-rich MC carbides (in this reference shown as NbC) with the increase of niobium content. (From Sawamoto, A. et al., *J. Jpn. Inst. Met.*, 49(6), 475, 1985.)   (*Continued*)

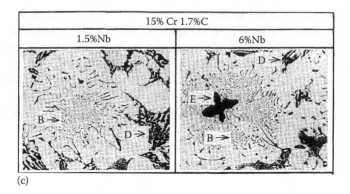

(c)

FIGURE 3.32 (Continued) Formation of MC carbides in high-C, high-alloy steels. (c) Microstructural examples for two compositions shown in (b). (Also from Sawamoto, A. et al., *J. Jpn. Inst. Met.*, 49(6), 475, 1985.), with fixed 1.7%C. B, eutectic MC; E, primary MC; D, eutectic M_7C_3.

MC particles of primary type, that is, no isolated particles precipitated in the liquid. When the V content increases to >1%, the thermodynamic driving force for MC precipitation also increases, which it starts to form directly in the liquid. In other words, the higher the V content, the longer the MC solidification interval, once this carbide is formed at a higher temperature and continues to grow until the eutectic temperature is reached and all the liquid converts to a eutectic morphology of either MC or M_2C eutectics. The solidification interval of primary MC determines not only the amount of primary MC but also the available time (for a constant cooling rate) during which they can freely grow in the liquid. Therefore, the solidification interval deeply influences the size of primary MC and can be changed by composition variations, especially the amount of V, Nb, C, and N (see Reference 44).* Another way to refine not only the primary but all carbides is the increase in cooling rate during solidification, discussed in the next section, which then decreases on average all the available time for solidification.

Similar to high-V tool steels, high Nb contents also lead to primary MC carbides, as shown in Figure 3.32b, where different amounts of Nb were added to high-Cr, high-C steel similar to D2 and D3. The increase in Nb shows a considerable increase in the total MC-type carbides, as expected, but the microstructures show that the amount of primary NbC is substantially formed only in the composition with higher Nb content. Not dissolving extensively in M_2C eutectics, Nb-rich MC carbides are

---

* Most low-diameter (less than ½ in.) high-speed steel tools are machined by grinding wheels after fully hardened. So the presence of large and hard particles is usually avoided, since they affect the manufacturing operation. Being very hard, the MC carbides confer high abrasive resistance, leading to higher resistance to be ground off and thus leading to overheating problems (cracks or hardness decrease) during the grinding of hardened tools. The refinement of carbides, especially the primary MC, is usually necessary for critical tools such as taps or special drills, as a way to keep the amount of carbide particles, important to tools performance, but to avoid excessive resistance during grinding.

then present either as a eutectic or in primary morphologies. NbC is also less soluble in austenite than V. So, as an overall effect, the amount of Nb necessary to form MC carbides is usually lower than the amount of V, and thus more MC carbides can be observed in high-Nb, high-C tool steels.

3.3.3.2 Variables Affecting Carbide Distributions

As shown in Section 3.3.2, the undissolved carbides are important to wear resistance and deleterious to toughness, and both effects depend mainly on the carbide distribution rather than the amount and type of carbides only. The final distribution, on the other hand, will depend on the initial as-cast microstructure, on the carbide eutectics of the primary carbides, and on the deformation of such microstructure during the forming operations, by hot-forging or hot-rolling.

On the first point, the *as-cast microstructure*, the types and basic morphology hardly ever change; that is, for a given composition, the carbides will maintain their types and the eutectic morphology. What vary, and very substantially, are the size of the dendrites and, consequently, the distribution of the carbide eutectics. The main parameter affecting the as-cast microstructure is the cooling rate during solidification: the faster the solidification ends, the finer the microstructure. One example of this correlation can be observed in Figure 3.33a, which compares the microstructures formed in different positions of the ingot, especially in relation to the distance to the ingot mold. In Figure 3.33b, it is clear that the regions closer to the mold present higher cooling rate, meaning faster solidification, and so finer microstructures (as shown in Figure 3.33a). The ultimate refinement of these microstructures is obtained in PM processes, converting the melt into small droplets, with submillimeter sizes, that solidify in contact with a high-pressure gas and thus very fast (a few orders of magnitude higher than the usual ingot casting). Then, the as-cast eutectics become extremely fine and lead to small and spherical carbides in the final products, very well distributed, as shown in Figures 3.25 and 3.27.

After casting, the carbide eutectic distribution is continuously "broken" by the applied deformation, during all *hot-forming* operations. Being composed of hard phases, the eutectics cannot follow the movement of the metal matrix volume during hot rolling and are continuously fragmented. Especially in small-diameter bars, such as high-speed steel for drills, the deformation of the microstructure may be very extensive, in the range of >1000 times in area. In Figure 3.34, this process of carbide fragmentation can be observed after two different deformation levels. It is interesting that the cellular shape of the eutectics is gradually changed, until a point where the carbides assume linear distributions, also named *carbide stringers*.

This process of deformation considerably improves the tool steel toughness, by reducing the carbide sizes and the sizes of carbide agglomerates (see Figure 3.28a). However, while the toughness increases, the properties become dependent on the direction, that is, the tool steel becomes nonisotropic (in any tool steel with aligned carbides). This occurs because the carbides are distributed in a totally different way when compared to the rolling direction (longitudinal) and the cross-sectional direction (transverse). One example of such anisotropy is shown in Figure 3.35, where a PM tool steel is tested directly after hot isotactic pressing (HIP), also known as the

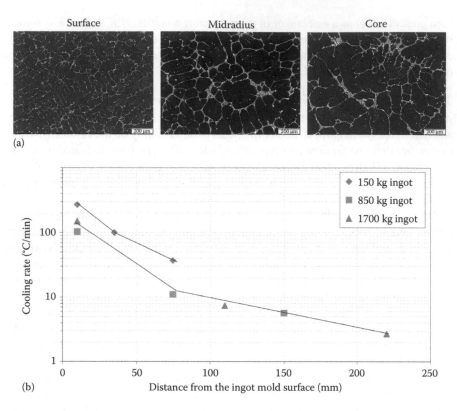

(a)

(b)

FIGURE 3.33 Example of as-cast microstructure of high-speed steels, showing (a) that the carbide networks get coarser in the inner regions, compared to the surface. (b) The calculated cooling rate for different ingot sizes, showing that, with small differences for different ingots, the cooling rate depends primarily on the distance to the ingot mold surface. (From Delgado, M.P. et al., High speed steel M2 solidification study in different ingots, *Proceedings of 64 Congresso Anual da ABM*, Belo Horizonte, Brazil, 2009, pp. 1–7.)

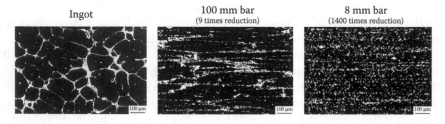

FIGURE 3.34 Variation after hot-rolling in the microstructure of a high-speed steel. Each picture shows the dimension after rolling and the area reduction. (From Mesquita, R.A. and Barbosa, C.A., *Mater. Sci. Forum*, 416/418, 235, 2003.)

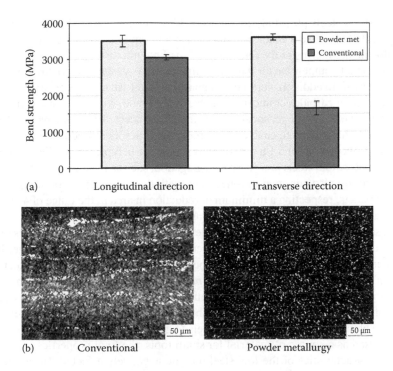

(a) Longitudinal direction   Transverse direction

(b) Conventional   Powder metallurgy

FIGURE 3.35 (a) Mechanical properties and (b) microstructural comparison between conventional and powder metallurgy high-speed steels. Observe that the homogeneity in mechanical properties is directly correlated with the homogeneity in carbide distributions in the microstructures. (From Mesquita, R.A. and Barbosa, C.A., *Mater. Sci. Forum*, 416/418, 235, 2003.)

HIP-ed condition, and compared to a tool steel with similar composition but produced by ingot casting and hot-rolling. Crack propagation dramatically changes in the conventionally rolled steel, being smaller when cracks propagate along aligned carbides (transverse testing) versus when cracks have to cross carbide agglomerates (longitudinal testing). So it is clearly observable in this example how the properties are homogeneous in the case of the PM tool steel, compared to the anisotropic behavior of the conventional steel. To enable this test, a large diameter was chosen (~100 mm or 4 in.), once the obtained specimens were enabled in both directions. Smaller diameters, if tested, would show higher toughness but also even higher anisotropy.*

Before ending this section, two implications may be derived from the refinement of the microstructure after casting and after hot deformation, by forging or rolling. The first is the balance between using *smaller* or *larger ingots* for cold work or high-speed tool steels. On one hand, larger ingots present a coarser microstructure but, on the other, the larger size enables higher amount of applied deformation until reaching the final dimension of the tool steel bar. One interesting study in this respect was done

---

* The actual toughness test of small diameter bars is not trivial for the transverse direction, because the length of the specimen is limited by the bar diameter and traditional methods (impact or bend test) usually require a minimum length of 50 mm (2 in).

for high-speed steels [48] studying ingots of 100, 1000, and 4000 kg, as summarized in Figure 3.36. It is clear that, for a reasonable amount of deformation, the ingot size is predominant in terms of the refinement of the final microstructure, because the hot-forming cannot compensate for the coarsening of the as-cast microstructure if very big ingots are used. In addition, very large ingots also lead to macro-segregation, meaning that areas full of carbides or unexpected phases are found, with sizes on the millimeter range. On the other hand, reasonably large ingots are necessary to achieve a minimum level of deformation and to enable good productivity in rolled products. So, in conclusion, using ingots of around 1 ton is a common practice in typical hot-rolled high-speed steels. For other tool steels less prone to segregation, such as D-series tool steels, ingots up to 4 tons are also common in rolled bars (with diameters up to 100 mm). And for forged products, respecting a minimum of reduction in area in the range of 4–10 times as well as choosing the smallest possible ingot is often the best strategy.

The second important point for tool steels containing undissolved carbides distributed in their microstructure is related to cutting smaller pieces from a bigger block. Very common in tool steel business is the fact that the steel distributors keep their stock in the form of blocks (also called *master blocks*) and cut from them smaller pieces to deliver to the tool makers. Although this is a common practice, the extrapolation of this practice to extreme sizes should be analyzed carefully. For example, if a large block of carbide-containing steels (e.g., D-series tool steel or a high-speed steel) is used as a source of material for small tools, the final properties or the heat-treating characteristics of the tool steel may be negatively affected. In Figure 3.37, an example is shown in this respect, simulating a cold work tool with small section size (thickness of ~20 mm) produced from a 22 mm thick rolled bar or from a piece cut from a larger (350×450 mm^2) block. It is clear from the toughness results that the mechanical properties are much higher (three to five times!) for the sample taken from the rolled bar. This means that if a small tool is made from tool steel taken from a rolled bar or from a large block, the final toughness will be affected. In other words, although the chemical composition, heat treatment, and even the hardness are the same, the fact that the tool steel came from a larger block deeply affects its performance, which is explained by the changes in the microstructures, also shown in Figure 3.37. While the carbide networks still show the cellular-like morphology in the large block, they are much finer and well distributed in the rolled bar. This microstructure refinement is common in small-section rolled bars of cold work or high-speed steels, resulting from both the higher degree of deformation and the fact that small rolled bars also start to form smaller ingots, with finer as-cast structures (as discussed earlier). In addition to the increase in toughness, a better distribution of carbides also leads to less distortion during heat treatment, which is another argument to avoid producing small tools from pieces cut from large blocks.

Therefore, while it is unavoidable that large tools contain a large amount of carbides (because they are made from tool steel taken from large blocks), the same can be avoided in small tools by not making them from pieces cut from larger blocks. And, as a basic recommendation, cold work and high-speed tools should be made with starting material as close as possible to the mill-delivered dimensions. By doing this, one can guarantee that the maximum possible microstructure refinement from the tool steel processing is achieved, enabling the best combination of the final mechanical properties.

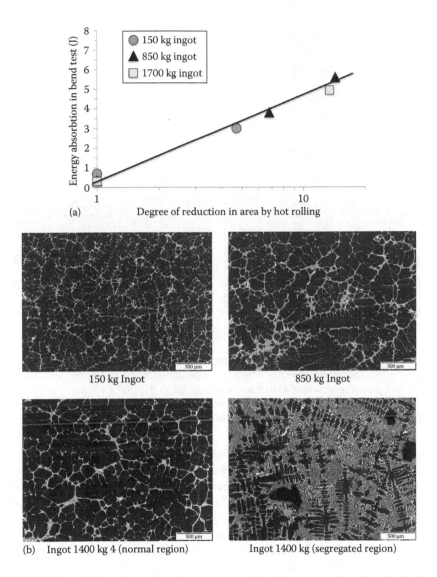

(a)

(b)   Ingot 1400 kg 4 (normal region)   Ingot 1400 kg (segregated region)

FIGURE 3.36 (a) Variation of toughness measured by the energy absorption during bend test (specimens of 5×7 mm^2 section size) versus the degree of reduction in hot rolling, for three different ingots. (b) Typical microstructures for half height the mid-radius position of each tested ingot, before hot rolling. Observe that, independent on the ingot size, more reduction leads to better toughness, as a result of the microstructure refinement. (From High speed steel M2 solidification study in different ingots and its effects in the microstructure and mechanical properties, Masters thesis, Universidade Federal de São Carlos, São Carlos, Brazil, 2010.)

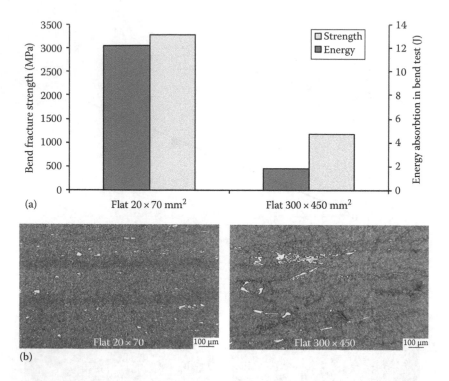

FIGURE 3.37 Comparison of samples taken from different flat bars of 0.8%C 8%Cr cold work tool steel. The smaller flat (20×70 mm^2) was produced from a smaller ingot (~2 tons) and with high degree of reduction of ~740 times, while the larger flat (300×450 mm^2) was produced from a larger ingot (~8 tons) and 7 times area reduction during hot forging. (a) Bend test results, comparing the toughness of both materials. (b) Typical microstrucutre for the tested pieces.

In tool steels that do not contain particles, such as hot work and plastic mold steels, this is not so critical and cutting from larger pieces is usually an acceptable practice. However, very extreme situations also should be avoided, not because of the particle distribution, usually nonexistent in plastic or hot work steels, but because of segregation aspects. When large tool steels blocks are necessary, they should be produced from large ingots (with 10 tons or more), which in turn means that the segregation during solidification starts to play an important role in the quality of the final microstructure. So smaller tools produced from smaller blocks also are less likely affected by segregation, and choosing the best cutting strategy is also important in hot work or plastic mold tool steels.

3.4 CLOSING REMARKS ON PHYSICAL METALLURGY OF TOOL STEELS

Although the next chapters of this book give detailed explanation of the properties and the microstructure of different types of tool steel, the understanding of the physical metallurgy aspects is crucial in fully understanding the involved metallurgical phenomena. In this chapter, those phenomena were explained at a reasonably good depth, which led to a full understanding of how the chemical composition and the manufacturing of a tool steel will influence its microstructure, which in turn will lead to different properties, while also considering the very important effect of heat treatment. Therefore, the physical metallurgy aspects will be reconsidered and referenced in the next chapters to explain the practical aspects (e.g., performance, failure, manufacturing methods, etc.) of tool steel and their applications.

REFERENCES

1. W.D. Callister, D.G. Rethwisch. *Materials Science and Engineering: An Introduction*, 4th edn. Hoboken, NJ: Wiley, 2010.
2. J.R.T. Branco, G. Krauss. Heat treatment and microstructure of tool steels for molds and dies. *Proceedings from Tool Materials for Molds and Dies*, Eds. G. Krauss and H. Nordberg, Golden, CO: Colorado School of Mines Press, 1987, pp. 94–117.
3. J.B. Bacalhau. Development of a 40 HRC plastic mold steel with high machinability. Master thesis of the Graduate Course in Aeronautic and Mechanical Engineering, São José dos Campos, Brazil: Instituto Tecnológico de Aeronáutica, 2012, 140pp. Original title: *"Desenvolvimento de aço para moldes plásticos com 40 HRC e elevada usinabilidade."*
4. NADCA #229/2006. Special quality die steel & heat treatment acceptance criteria for die casting dies. Holbrook Wheeling, IL: North American Die Casting Association—NADCA, 2006, 33pp.
5. L.C.G. Silva Jr., E.N. Ranzani, C.A. Barbosa, R.A. Mesquita, A. Sokolowski, H.F. Fasolin, A. Murari, M.A.S. Nogueira. Implementation of a rough rolling for high alloy steels at villares metals S.A., 2004. *Proceedings of 59 Congresso Anual da ABM (59th Annual Conference of Brazilian Metallurgical and Materials Society)*, São Paulo, Brazil, 2004, pp. 297–305 (Original in Portuguese, title: Implantação do Laminador de Barras Grossas (Billet Mill) da Villares Metals S.A).
6. S. Wilmes, K.P. Burns. Comparison of the toughness of hot work tool steel from different manufacturing processes with regard to the use in die casting dies. *Gießerei*, 76(24), 835–842, 1989. (Original in German, title: Vergleich der Zähigkeit von Warmarbeitsstahl unterschiedlicher Herstellverfahren im Hinblick auf die Verwendung für Druckgießformen).
7. G. Roberts, G. Krauss, R. Kennedy. *Tool Steels*, 5th edn. Materials Park, OH: ASM International, 1998, pp. 75–77.
8. H. Chandler. *Tool Steels, Heat Treaters Guide: Practice and Procedures for Irons and Steels*. Materials Park, OH: ASM International, 1995, pp. 517–669.
9. J. Ohlich, A. Rose, P. Wiest. *Atlas zur Wärmebehandlung der Stähle, Bd. 3, ZTA-Schaubilder*. Düsseldorf, Germany: Verlag Stahleisen mbH, 1973.
10. Uddeholm Tooling. Technical catalogue for tool steels. Available online at: http://www.uddeholm.com/files/. Accessed December 2015.
11. Böhler Edelstahl. Technical catalogue for tool steels. Available online at: http://www.bohler-edelstahl.com/english/1853_ENG_HTML.php. Accessed December 2015.
12. Villares Metals. Technical catalogue for tool steels. Available online at: http://www.villaresmetals.com.br/pt/Produtos/Acos-Ferramenta. Accessed December 2015.
13. A.H. Rosenstein. Interpretation of stress relaxation data for stress-relief application. *Journal of Materials*, 6(2), 265–281, 1971.
14. F.D. Richardson. The thermodynamics of metallurgical carbides and of carbon in iron. *Journal of the Iron and Steel Institute*, 175, 33–51, 1953.

15. H. Kwon, W. Kim, J. Kim. Stability domains of NbC and Nb(CN) during carbothermal reduction of niobium oxide. *Journal of the American Ceramic Society*, 98, 315–319, 2015.

16. D.A. Porter, K.E. Easterling. *Phase Transformations in Metals and Alloys*. Boca Raton, FL: CRC Press/Taylor & Francis Group, 2004.

17. F. Kayser, M. Cohen. Carbides in high speed steels: Their nature and quantity. *Metal Progress*, 61(6), 79, 1952.

18. G.A. Roberts, R.A. Cary, *Tool Steels*, 4th edn. Materials Park, OH: American Society for Metals, 1980, pp. 645–653.

19. G. Bandel, H.C. Haumer. Determining by calculation the full-hardening characteristics of long-section forgings. *Stahl und Eisen*, 84(15), 1946, 932–946 (in German).

20. K.E. Thelning. *Steel and Its Heat Treatment. Borfors Handbook*, Boston, MA: Butterworths, 1975, pp. 169–176.

21. L.C.F. Canale, R.A. Mesquita, G.E. Totten (Eds.). Failure analysis of heat treated steel components. *Failure Analysis in Tool Steels*, Eds. R.A. Mesquita, C.A. Barbosa, Materials Park, OH: American Society for Metals (ASM), 2008, vol. 1, Chapter 11, pp. 311–350.

22. M.L. Schmidt. Effect of austenitizing temperature on laboratory treated and large section sizes of H-13 tool steel. *Tool Materials for Molds and Dies: Application and Performance*, St. Charles, IL, Eds. G. Krauss, H. Nordberg, Golden, CO: Colorado School of Mines (CSM) Press Center, 1987, pp. 118–164.

23. B.L. Averbach, S.A. Kulin, M. Cohen. The effect of plastic deformation on solid reactions, Part II: The effect of applied stress on the martensite reaction, *Cold Working of Metals,* American Society for Metals, 1949. A seminar on the cold working of metals held during the *Thirtieth National Metal Congress and Exposition*, Philadelphia, PA, October 23–29, 1948.

24. R. Ebner, H. Leitner, F. Jeglitsch, D. Caliskanoglu. Methods of property oriented tool steel design. *Proceedings of the Fifth International Tooling Conference*, Leoben, Austria, 1999, pp. 3–24.

25. J.H. Hollomon, L.D. Jaffe. Time-temperature relations in tempering steel. *Transactions of the Metallurgical Society of AIME*, 162, 223–249, 1945.

26. W. Crafts, J.L. Lamont. Secondary hardening of tempered martensitic alloy steel. *Transactions of TMS-AIME*, 180, 471, 1949.

27. R.A. Mesquita, H.J. Kestenbach. Influence of silicon on secondary hardening of 5 wt.% Cr steels. *Materials Science & Engineering. A, Structural Materials: Properties, Microstructure and Processing*, 566, 970–973, 2012.

28. B.R. Banerjee. Embrittlement of high-strength tempered alloy martensites. *Journal of the Iron and Steel Institute*, 203, 166–174, February 1965.

29. H. Brandis, E. Haberling, H.H. Weigard. Metallurgical aspects of carbides in high speed steels. *Processing and Properties of High Speed Tool Steels*, Eds. M.G.H. Wells, L.W. Lherbier, Warrendale, PA: TMS-AIME, 1980, pp. 1–18.

30. L.P. Tarasov. The microhardness of carbides in tool steels. *Metal Progress*, 54(6), 846, 1948.

31. E. Elsen, G. Elsen, M. Markworth. Vanadium alloyed high-speed steels. *Metall: internationale Zeitschrift für Technik und Wirtschaft*, Vol. 19, pp. 334–345, 1965 (original in German)

32. G.B. Brook, J.M.G. Crompton, Fulmer Report R 319/4, Fulmer Research Institute, 1971. Apud: S. Karagoz, H. Fischmeister, *Metallurgical Transactions A*, 19Z, 1935–1941, 1988.

33. R.A. Mesquita, C.A. Barbosa. An evaluation of wear and toughness in cold work tool steels. *Tecnologia em Metalurgia e Materiais*, 2, 12–18, 2005. (In Portuguese, original title: *"Uma avaliação das propriedades de desgaste e tenacidade em aços para trabalho a frio"*).

34. R. Scheneider. The performance of spray-formed tool steels in comparison to conventional route material. *Proceedings of the Sixth International Tooling Conference—The Use of Tool Steels: Experience and Research*, Eds. J. Bergstrom, G. Fredriksson, M. Johansson, O. Kotik, F. Thuvander, Karlstad University, September 10–13, 2002, pp. 1111–1124.

35. I. Schruff, V. Schüler. Advanced tool steels produced via spray forming. *Proceedings of the Sixth International Tooling Conference—The Use of Tool Steels: Experience and Research*, Eds. J. Bergstrom, G. Fredriksson, M. Johansson, O. Kotik, F. Thuvander, Karlstad University, September 10–13, 2002, pp. 1159–1179.

36. R.A. Mesquita, C.A. Barbosa. Powder metallurgy and spray formed 10%V cold work tool steels, a comparison. *Proceedings of Euro PM 2004 (Powder Metallurgy World Congress & Exhibition)*, 5, 53–64, 2004.

37. O. Grinder. PM HSS and tool steels—Present state of the art and development trends. *Proceedings of the Fifth International Conference on Tooling*, Loeben, Austria, 1999, pp. 39–47.

38. G. Hoyle et al. A modified bend test for hardened tool steels. *Journal of the Iron and Steel Institute*, 191, 44–55, 1959.

39. E. Horn, H. Brandis, *DEW Technische Berichte*, 11, 147–154, 1971. Apud: G. Roberts, G. Krauss, R. Kennedy. *Tool Steels*, 5th edn. Materials Park, OH: ASM International, 1998, pp. 257.

40. K. Bungardt, E. Kunze, E. Horn. Investigation of the structure of the iron-chromium-carbon system. *Archiv fur das Eisenhuttenwesen*, 29, 193, 1958.

41. H. Berns. Comparison of wear resistant MMC and white cast iron. *Wear*, 254, 47–54, 2003.

42. R.A. Mesquita, C.A. Barbosa. Hard alloys with lean composition. Patent: US 8,168,009, Priority: 2006.

43. R.A. Mesquita, C.A. Barbosa, Spray forming high speed steel—Properties and processing. *Materials Science & Engineering A*, 384, 87–95, 2004.

44. H. Mizuno, K. Sudoh, T. Yanagisawa. The influence of alloying elements on the morphology of MC primary carbide precipitation in Mo-type high speed tool steel. *Denki Seiko*, 55(4), 225–236, November 1984.

45. A. Sawamoto, K. Ogi, K. Matsuda. Solidification structures of Fe-C-Cr-Nb alloys. *Journal of Japan Institute for Metals*, 49(6), 475–482, 1985.

46. M.P. Delgado, R.A. Mesquita, C.A. Barbosa. High speed steel M2 solidification study in different ingots. *Proceedings of 64 Congresso Anual da ABM*, Belo Horizonte, Brazil, 2009, pp. 1–7. (In Portuguese, with original title: "Microestrutura e Propriedades Mecanicas do Aço Rapido M2, Produzido a Partir de Lingotes de Diferentes Dimensões").

47. R.A. Mesquita, C.A. Barbosa. Evaluation of as-HIPped PM high speed steel for production of large-diameter cutting tools. *Materials Science Forum*, 416/418, 235–240, 2003.

48. M.P. Delgado. High speed steel M2 solidification study in different ingots and its effects in the microstructure and mechanical properties. Master thesis, Universidade Federal de São Carlos, São Carlos, Brazil, 2010 (Original in Portuguese: "*Estudo da Solidificação do Aço Rápido M2 para diferente lingotes e seus efeitos na estrutura e propriedades mecânicas finais*").

4 Hot Work Tool Steels

4.1 INTRODUCTION TO HOT FORMING TOOLS

One of the most important properties of metals is their ability to be shaped or formed, enabling the production of parts with a variety of shapes and forms. However, the ability to plastically deform metals is limited at room temperature. Once a strain is applied, the number of dislocations increases, causing strain hardening and the reduction of ductility; and with continued application of strain, this will eventually lead to cracks in the formed part. One way to increase the amount of ductility is by avoiding strain hardening, by means of reducing the number of accumulated dislocations in the microstructure, and which in turn is possible by increasing the temperature. Above a certain temperature, usually above 600°C in steels, several microstructure mechanics causing softening may take place, promoting dislocation arrangements (restoring) or creating new grains (recrystallization). Through both mechanisms, the ductility increases and then the metal can be deformed to a much higher extent. At a higher temperature, dislocations will move faster and overcome easily their obstacles, meaning that the strength of the metal being hot-formed then decreases. The net effect is that, at high temperatures, metals can be deformed to a much larger extent and with lower forces. And softening is faster and more effective once the temperature is increased, for instance above 1000°C (1830°F) for steels.

Therefore, several hot-forming processes have been developed such that metals, mainly steels and aluminum alloys, are able to be shaped into complex forms and with high-productivity industrial processes. The main hot-forming processes used today are forging, rolling, and extrusion. With exception to hot-rolling,* all the other processes involve hot work tool steels as the "mold" for the formed parts, but the terminology "dies" or the generic term "tools" is more common in the industry, reserving the term "molding" for processes involving shaping from a liquid, nonmetallic material, such as plastics.

In addition to their use in forming or shaping, hot work tool steels are also employed as casting dies for low-melting-point metals, under high pressure or only under gravity. In particular, for aluminum cast parts, pressure casting is very popular for producing highly complex parts with good quality and in a short time, named "high-pressure die-casting" or often only "die-casting." For steel casting or cast iron,

---

* Hot-rolling is usually of a different characteristic than the other hot-forming process, because it involves simpler geometries and fast processing. The presence of scale and the high speed between the rolled piece and the rolls make give rise to high abrasion and the short contact with high cooling does not lead to as extensive heating on the rolled surface as in the other hot work tools. Therefore, white cast irons or high-Cr, high-carbon forged steels are commonly used for rolls, by offering high wear resistance with moderate hot strength. When a combination of wear resistance and high temperature resistance is necessary, the rolls employ high-speed steels, which, similar to hot work steels, present strong secondary hardening and high temperature strength, but also present undissolved carbides for abrasion resistance.

the temperatures are extremely high (from 1100°C to 1500°C/2000°F to 2700°F) and usually the molds are made of sand or other ceramic materials. But, for aluminum, the temperature of the liquid is usually ~700°C (1300°F) and steel may be used as "mold," because the tool surface, due to the short contact time, reaches <650°C (1200°F). Under these conditions, precipitation hardening of hot work tool steel is

(a)   (b)

FIGURE 4.1   Examples of hot-working dies, for application in (a) hot forging, used to produce crankshaft, and (b) die-casting, to cast an aluminum gearbox cover. (From Canale, L.C.F. et al. (eds.), Failure analysis of heat treated steel components, in: *Failure Analysis in Tool Steels*, Mesquita, R.A. and Barbosa, C.A., eds., American Society for Metals (ASM), Materials Park, OH, 2008, vol. 1. Chapter 11, pp. 311–350.)

TABLE 4.1
Chemical Composition of the Main Tool Steels Applied in Industry Showing the Commonly Observed Target Values

Hot Work Tool Steels			Most Common Chemical Composition (wt.%)							
ASTM	EN/DIN	UNS	C	Si	Mn	Cr	V	W	Mo	Other
6F3 mod	1.2714	T61206	0.56	0.3	0.8	1.1	0.1	—	0.5	Ni = 1.7
H10	1.2365 mod	T20810	0.34	1.0	0.3	3.2	0.4	—	2.5	—
H10 mod	1.2367	—	0.38	0.3	0.3	5.0	0.5	—	3.0	—
H11	1.2343	T20811	0.36	1.0	0.3	5.0	0.4	—	1.3	—
Low Si H11	1.2343 mod	—	0.36	0.3	0.3	5.0	0.4	—	1.3	P < 0.015
H12	1.2606	T20812	0.36	1.0	0.3	5.0	0.3	1.5	1.5	—
H13	1.2344	T20813	0.38	1.0	0.3	5.0	0.9	—	1.3	—
H19	1.2678	T20819	0.40	0.3	0.3	4.2	2.0	4.2	0.2	Co = 4.25
H21	1.2581	T20821	0.32	0.3	0.3	3.5	0.4	9.0	—	—

Note: The specific limits and ranges are given by the standards ASTM A600 and Euro Norm ISO 4957 (old DIN 17350 and EN 10027), according to References 2 and 3. Throughout this chapter, the DIN numbers are used.

enough to avoid softening, and thus aluminum die-casting is today a large field of application for hot work tool steels.

Examples of tools for both processes, namely forging and die-casting, are shown in Figure 4.1. It is clear from both pictures that these tools are of complex and of large size, which leads to high production costs, in terms of machining and heat treatment, in addition to the cost of the tool steel, which in many cases is highly alloyed and so involves high unit cost. Therefore, several hot work dies may require long and expensive production processes, with the final cost varying from tens to hundreds of thousands of dollars.

To achieve the necessary properties and performance in all these diverse hot-forming operations, a series of tool steels have been developed since the 1940s, of special importance being the H-series of ASTM or similar other standards. The main grades used in hot work tooling are discussed in this chapter, and their chemical compositions are shown in Table 4.1.

4.2 METALLURGICAL PROPERTIES OF MAIN HOT WORK TOOL STEELS

To perform well during the temperatures developed in the hot-forming processes, hot work tool steels have a large combination of metallurgical properties, such as high-temperature strength, resistance to softening (tempering resistance), resistance to cracking under impact and fatigue—including fatigue stresses emerging from temperature variations on die surface (thermal fatigue)—thermal conductivity, adequate response to heat treatment and surface modification treatments (mainly nitriding), and wear resistance at high temperature, either in the heat-treated or nitrided condition.

In the present section, we discuss all these properties, falling under three main groups: (1) properties related to the strength at high temperatures and tempering resistance, the last term concerning the stability of strength during hot working; (2) toughness, in the widest perspective of this property, concerning resistance to all kinds of cracks and fracture mechanisms; and (3) resistance against wear, especially abrasive wear, which is primarily determined by the high-temperature surface hardness, affected either by the tempering resistance or by nitriding or other surface treatments.

4.2.1 HIGH-TEMPERATURE STRENGTH AND TEMPERING RESISTANCE

Perhaps the first property that comes to mind when one thinks of hot work tool steels is the hot strength, that is, the ability to withstand loads at high temperature. Before going ahead with all explanations, it is relevant to discuss what "high temperature" means here. A precise definition of hot work temperatures is not found in the literature or standards. The range 450°C–500°C (840°F–930°F) is typical for billets in aluminum extrusion, which is a continuous process and when the temperature of the die surface and near-surface is in equilibrium with the billet. Hot-forging temperatures usually exceed this range by a large extent, achieving easily 1000°C–1200°C (1830°F–2190°F), but because of the short contact time and the application of lubricants/coolants, regions a few millimeters below the surface are also in range of ~600°C (1110°F) or less. In permanent-mold casting, or die-casting, the industry usually uses cast aluminum or magnesium, with casting temperatures of ~700°C (1290°F) but also with short contact times. In an overall view of the industrial practice, hot work tools may then be defined when the formed piece is around or above 500°C, with the dies and especially the dies surface heated to such temperatures.

The first step to understand the behavior of tool steels used in hot work tooling is then to measure and compare the high-temperature strength curves. Some examples of such curves are shown in Figure 4.2 for different types of steel. The first observation is that, for all steels, the strength continuously decreases with the increase in temperature, which may be due to two main factors. The first is that the dislocations' ability to move and overcome barriers is higher at higher temperatures. But the second and very important reason is given by the strengthening mechanisms of tempered martensite. All shown steels are hardened and tempered, the strength being only initially determined by the martensite hardness. As explained in Section 3.2.2.3, the increase in temperature leads to tempering effects in a martensitic microstructure, where carbon moves out from solid solution in martensite and form carbides. When iron carbides (cementite) are formed, there is a continuous decrease in the strength (tempered hardness), but if alloy carbides are present, hardness may be sustained at higher levels at higher temperatures and for longer times. This then explains the difference in strength of the steels shown in Figure 4.2, which is caused by the different amounts of alloying elements with the ability to form fine and well-distributed secondary carbides, which are mainly V, Mo, and W, and to a lower extent Cr. So, the increase in

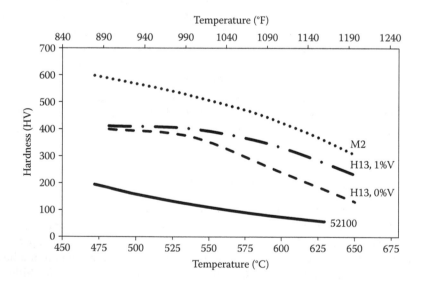

FIGURE 4.2 Hot hardness of selected alloyed steels: 52100 (1%C, 1.4%Cr); H13, 0%V (0.33%C, 5%Cr, 0.8%Si, 1.35%Mo); H13, 1%V (0.33%C, 5%Cr, 0.8%Si, 1.3%Mo, 1%V); M2: (0.9%C, 5%Mo, 6%W, 2%V). (The charts for M2 and 52100 were built with data extracted from Jatczak, C.F., *Metal Prog.*, 70, 1978; H13 data from Roberts, G.A. and Cary, R.A., *Tool Steels*, 4th edn., American Society for Metals, Beachwood, OH, 1980, pp. 645–653.)

strength from the low-alloy 52100 steel to H13 is caused by the differences in Cr and Mo, and finally the addition of high Mo and W explains the difference between M2 and high V H13.*

Although using the data of high-temperature strength may be important in specific cases, knowing that the main mechanism is the precipitation hardening of alloy carbides may lead us to simpler and more relevant data. In this sense, steels with higher amount of and more stable precipitates will show better behavior in terms of hot strength if these carbides persist for longer times, blocking the movement of dislocations. Time is in fact a very important variable in hot work tooling, since tools and dies are used to produce typically thousands to hundreds of thousands of parts. Although the contact with the hot piece may be short (usually a few seconds), the repetition for this number of produced parts leads to cumulative heating over a time range of hours. So, the tool steel must withstand the high temperature loads but also must not be strongly affected by the heating during the process. A second comparison graph is built in this sense (Figure 4.3), giving the hardness, measured at room temperature, but after exposure for different times to high temperature. While this is a very simple diagram in terms of experiment is also a very convenient way to measure how stable the hardness is for a given tool steel and also to compare different types of steels. In fact, the exposure to higher temperature causes the tempering

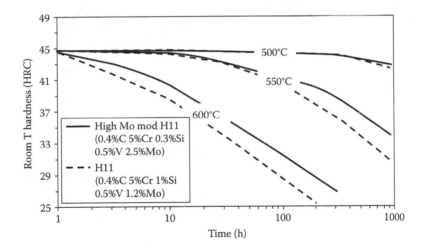

FIGURE 4.3 Decrease in hardness (measured at room temperature) after exposure to high temperatures, for two different tool steels. Observe the increase in the resistance to softening, properly called *tempering resistance*, when the Mo content is doubled. (Drawn with data from Uddeholm Tooling, Technical catalogue for tool steels, Available online at: http://www. uddeholm.com/files/, Accessed December 2015, for steels Dievar and Orvar.)

---

* In M2 tool steel, a substantial part of the alloying elements is formed by undissolved carbides, which are important for wear resistance (Section 3.3.2.1) but not for the tempered hardness. Nevertheless, about 3%Mo, 3%W, and 1%V are dissolved after hardening, which leads to very high hardness and tempering resistance for this grade, superior to those of H13 (see Figure 4.2).

of the microstructure, and the hardness will always decrease, for a given time and temperature, but by an amount that will depend on the ability of the tool steel to avoid the loss in hardness by the applied tempering. This brings us to a very important property of hot work tool steels, known as *tempering resistance*, which is the ability to resist hardness decrease after exposure to high temperatures. And, because of the continuous heating of tools during use, tempering resistance determines the performance of hot work tool steels in many situations, especially when concerning deformation or wear at high temperature.

Therefore, the hot strength of tool steels primarily depends on their tempering resistance, which may be evaluated in simple charts as in Figure 4.3, but may also

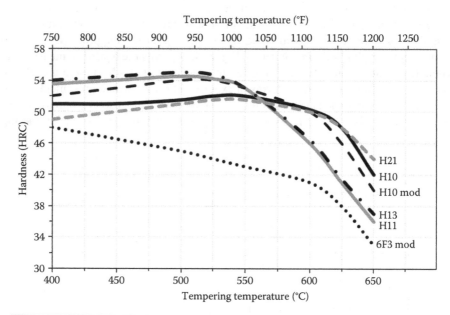

ASTM	DIN	Hardening	Hardness (HRC)							
				Tempering, 2 × 2 h						
Hot work tool steels			As Quench	400°C (750°F)	450°C (840°F)	500°C (930°F)	525°C (980°F)	550°C (1020°F)	600°C (1110°F)	625°C (1160°F)
6F3 mod	1.2714	900°C (1650°F)	59	48	47	45	44	43	41	38
H10	1.2365	1020°C (1870°F)	52	51	51	52	52	52	50	48
H10 mod	1.2367	1030°C (1885°F)	56	52	53	54	54	53	50	46
H11	1.2343	1010°C (1850°F)	55	53	54	54	54	54	46	41
Low Si H11	1.2343 mod	1010°C (1850°F)	53	52	52	53	52	51	46	42
H13	1.2344	1020°C (1870°F)	55	54	54	55	55	53	47	42
H21	1.2581	1100°C (2010°F)	50	49	50	51	52	51	50	48

FIGURE 4.4 Tempering diagrams for the most common hot work tool steels, showing hardness after two times tempering for 2 h. Data from brochures of traditional tool steel producers. (From Uddeholm Tooling, Technical catalogue for tool steels, Available online at: http://www.uddeholm.com/files/, Accessed December 2015; Böhler Edelstahl, Technical catalogue for tool steels, Available online at: http://www.bohler-edelstahl.com/english/1853_ENG_HTML.php, Accessed December 2015; Villares Metals, Technical catalogue for tool steels, Available online at: http://www.villaresmetals.com.br/pt/Produtos/Acos-Ferramenta, Accessed December 2015.)

be qualitatively compared using the steels' tempering diagrams. Because tempering depends on both the time and temperature and that this correlation can be given by the Hollomon–Jaffe parameter ($P = T$ [20 + log t], with T in K and t in hours, explained in Section 3.2.2.3) [7], the tempering resistance of different steels can also be compared by the tempering charts of those steels. In Figure 4.4, the main hot work tool steels and their typical tempering charts arc shown, and one can immediately identify that the curves that are more to the right represent tool steels with higher tempering resistance. For example, for the 600°C condition (which is actually 2 × 2 = 4 h tempering at 600°C), 6F3 would have a hardness of 41 HRC, while H11 still has 46 HRC and H10 ~50 HRC. This means that, 6F3 has less tempering resistance than H11 and that H10 has more tempering resistance than both. And by simple calculations using the Hollomon–Jaffe parameter, any time and temperature combination may be converted to an equivalent condition for the 4-h total tempering times of the tempering charts. In summary, the earlier discussion shows that tempering charts, although commonly used to determine the adequate tempering temperature for a given hardness, may also be used as a comparison method for the stability of hot strength, or tempering resistance, of hot work tool steels.

One example of the tempering resistance and its application is shown in Figure 4.5, which describes a hot-forming punch used for high-speed forging (e.g., in Hatebur

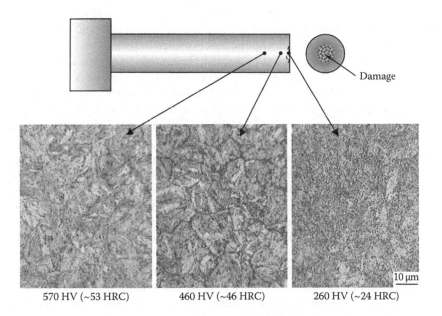

570 HV (~53 HRC)   460 HV (~46 HRC)   260 HV (~24 HRC)

FIGURE 4.5 Example of a forging punch that was severely tempered in service, leading to wear in the working area (tip), pointed as "damage." Observe that the heat leads to changes in the microstructure and decreases the hardness, as a result of tempering during service. The steel is a modified H10 (DIN 1.2365). (From Canale, L.C.F. et al. (eds.), Failure analysis of heat treated steel components, in: *Failure Analysis in Tool Steels*, Mesquita, R.A. and Barbosa, C.A., eds., American Society for Metals (ASM), Materials Park, OH, 2008, vol. 1, Chapter 11, pp. 311–350.)

presses). This case shows a steel with very high tempering resistance (similar to DIN 1.2365), which shows wear due to hardness decrease after the production of thousands of parts. The heating in the surface areas was so intense that hardness reduced to a very low level. Two solutions were suggested in this case, one related to the usage of a tool steel with an even higher tempering resistance (e.g., a Co-alloyed high-speed steel), or, the second, the increase in the efficiency of cooling to reduce heating of the surface areas (see Reference 1).

4.2.2 Toughness

Although withstanding the stresses at high temperature is the primary role of hot work tools under normal conditions, there are situations where tools may fail for several reasons not involving the lack of strength or wear resistance. Examples of those failures are often related to cracks, which may vary in size, number, and intensity, such as distributed thermal cracks, often called *heat checking*; mechanical cracks caused by mechanical fatigue; gross cracking, usually deep and large; or even catastrophic fractures that can occur all over the tool and usually cause the immediate interruption of die usage. Examples of all these factors are shown in more detail in Section 4.3 when their applications are described. Understanding the cracking of high-strength steels may be a very complicated task, especially when specific calculations or predictions are involved. However, the issue can always be simplified if fracture-related issues are considered as a matter of initiation and propagation of cracks and the microstructural features are correlated to both.

The first important correlation to cracking failure and toughness of tool steels is the relation between hardness and toughness, especially in hot work tool steels. As can be seen in the diagrams of Section 4.2.1, hot work steels may reach hardness levels in the range of 50–55 HRC but are often not used at these levels. The reason is that the maximum hardness is limited by the decrease in toughness and the failures that are likely to occur when the toughness is too low. For a balance between hardness and toughness, graphs like those shown in Figure 4.6 may be used. Specifically for the H13 steel charts shown in this figure, one can see that, when the hardness increases, toughness decreases, and vice versa. It is actually advised that the maximum hardness range should be avoided because of the extreme decrease in toughness. In addition, it can be seen that the retained austenite is reduced after tempering above 550°C (1020°F), leading to less dimensional changes and the problems related to retained austenite.

In addition to decreased toughness, tempering at lower temperatures may also be not effective for the hot strength when tempering resistance is considered. For example, if the surface reaches high temperatures during use, the hardness will naturally decrease, regardless of the initial hardness. In other words, if low temperatures are used for the tool tempering treatment, regions on the tool surface will be tempered during service, and the initial hardness will have only a small or no effect in the final hot strength. As a final result, using a low tempering temperature may just result in low core toughness, while the surface will be at lower hardness anyway. So, in hot work tooling, it is always recommended to consider tempering during service and tune the toughness accordingly, to avoid premature cracks. In general, the common

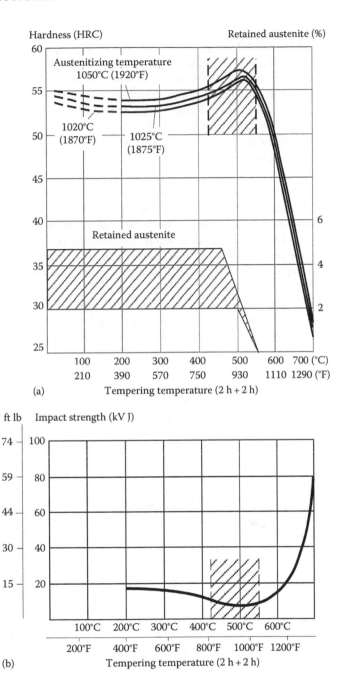

FIGURE 4.6 Correlation between (a) hardness and (b) toughness in H13 hot work tool steel. Observe that the region near ~500°C (930°F) is to be avoided because the very high hardness and, as a consequence, very low toughness. (From Uddeholm Tooling, Technical catalogue for tool steels, Available online at: http://www.uddeholm.com/files/, Accessed December 2015, steel Orvar Supreme.)

hardness levels are between 42 and 50 HRC, with tempering in the range 580°C to 630°C (1080°F to 1180°F), leading to adequate hot strength, good stability in terms of tempering resistance, and adequate toughness.

In terms of microstructure, the increase in hardness and decrease in toughness are directly related to specifically the resistance to crack propagation or fracture toughness. In most steels (and actually most materials), the fracture toughness decreases with the increase of strength, which means that the applied stresses cannot be accommodated and then it is energetically more favorable for a crack to propagate than to undergo plastic deformation.

All the earlier discussion was based on room-temperature toughness, but for all steels toughness is strongly affected by the temperature of the environment. This is related to the traditional brittle transition temperature, which is of special importance for medium-carbon steels such as hot work tool steels, with a carbon content between 0.3% and 0.6% [2, 3]. In those steels, toughness is strongly affected by the temperature. Examples of a few hot work tool steels are shown in Figures 4.7 and 4.8, where an increase of more than 100% is often observed between room temperature and ~300°C/400°C (570°F/750°F). This is one of the reasons why preheating to this range is a common and recommended practice before tools are put to use. The second reason is to decrease the temperature gradient between the surface and the regions underneath, which decreases the thermal stresses and the likelihood of

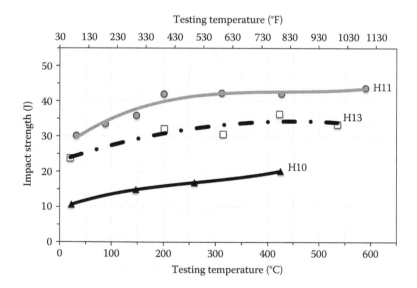

FIGURE 4.7   Influence of the environmental temperature on the toughness of hot work tool steel. Observe the substantial increase (from 50% to 100%) in toughness when the testing temperature increases from room temperature to ~300°C/400°C (~600°F/800°F). This justifies that tools be preheated before use. Impact strength refers to Charpy V. Starting hardness is 56 HRC for H10 and 52 HRC for H11 and H13. (Drawn with data from Hamaker, J.C., *Met. Prog.*, Vol. 70, p. 93, December 1956.)

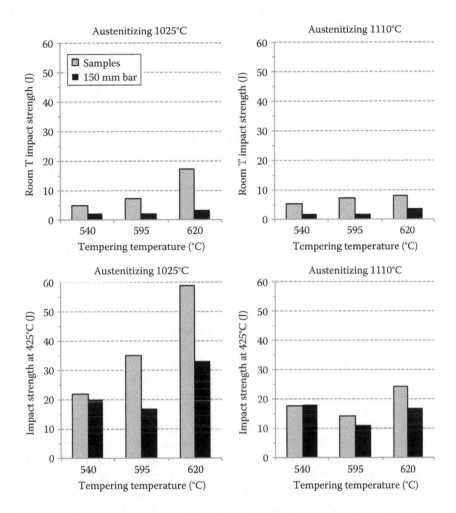

FIGURE 4.8 Influence of specimen size on the toughness after quenching and tempering, for two austenitizing temperatures (1025°C and 1110°C)—charts on the left and right, respectively—and two testing temperatures: room temperature toughness and toughness at 425°C—upper or bottom charts, respectively. Samples refer to about 10 mm section size specimen, while the 150 mm bar refers to specimens slowly quenched, in a condition that represents the core of a 150 mm round bar after oil quench. Impact strength refers to Charpy V. (Charts made with data from Schmidt, M.L., Effect of austenitizing temperature on laboratory treated and large section sizes of H-13 tool steels, *Proceedings of Tool Materials for Molds and Dies*, Krauss, G. and Nordberg, H., eds., Colorado School of Mines Press, Golden, CO, 1987, pp. 118–164.)

thermal cracks. Being carried out far below the tempering temperature, preheating does not affect the hardness by tempering effects.*

Another factor to consider on the toughness is the *quench embrittlement* phenomenon, as described in Section 3.2.2.2. This is especially important in hot work tool steels because these tools are often large (section sizes with hundreds to thousands of millimeters, or up to 40 inches) and also because of the alloying elements contained and their ability to form carbides. During quenching, these alloying elements, mainly V and Mo, may precipitate in the grain boundaries of austenite before it is transformed to martensite or bainite. The intensity of this precipitation will depend on the cooling rate, with slower quenching causing a higher amount of precipitates. These precipitates usually lead to a dramatic decrease in toughness, as exemplified in Figure 4.8. It is clear in those charts how toughness decreases from the quenched specimens to 150 mm bars, which is directly related to the precipitation of carbides in grain boundaries and to quench embrittlement. The differences are very substantial, more than 3–4 times smaller toughness values for the slowly quenched specimens. Although of importance of this effect, quench embrittlement is often neglected. Only recently (last 10 years) was the concern of a minimum quench rate introduced in the die-casting recommendation of NADCA [11], and it is often not recognized for other applications such as forging dies, which may also be of large size.

In Figure 4.8, it is also seen that the effect continues to be important at high temperatures (425°C), that is, after the tools are preheated. This shows that the toughness is further reduced after higher hardening temperatures are used, which intensifies the grain coarsening of austenite but also the precipitation during quenching, leading to further quench embrittlement. The worst combination is then the use of a high temperature and slow cooling, which can lead to a strong decrease in toughness, in some cases by a factor of almost 10, as shown under some conditions in Figure 4.8.

Several implications of quench embrittlement, based on the previously discussed mechanism, can be detailed: (1) Do not exaggerate on the austenitizing temperature and time, avoiding austenite grain growth, and excessive dissolution (and future precipitation) of alloy carbides. (2) Employ a fast cooling rate, even if the hardness of tool steel is achieved with a slower quenching.† (3) It is often not possible to measure the toughness after the die is heat-treated, so the procedure of quenching must be controlled to achieve the necessary cooling rates to minimize quenching embrittlement. In some cases, coupons may be connected (spot welded) to the die to evaluate the toughness

---

* In fact, it is a common rule that the hardness is not affected when steel is exposed to a temperature at least 30°C–50°C (55°F–90°F) below the tempering temperature.

† This also involves not using prehardened blocks to toughness sensitive applications, such as die-casting dies, and always heat treated dies in the premachined condition. This guarantees that the working surface of the final die is quenched in the fast cooling condition, which do not occur in hardened blocks, since the die-surface is placed in the interior parts of the block, where the quenching was performed with a smaller cooling rate and consequently with higher quench embrittlement. The same effect is less critical in low-alloyed hot work or mold steels, as explained in Section 3.2.2.2 of Chapter 3.

after quenching, which is actually recommended by NADCA [11].* (4) The application of high quenching speeds may not be trivial in vacuum-treated dies, with nitrogen gas quench, because a very high cooling rate may lead to cracks or distortion. But a balance between quench embrittlement and distortion must be respected; however, because distortion is easily observed and quench embrittlement is not, the second is often neglected to avoid the first. (5) Lastly, and in general, hardness and lack of distortion should not be considered as the only parameters for the quality of the heat treatment, but the final toughness and the quench embrittlement must also taken into account.

In terms of specific tool steels and the effect of the manufacturing conditions on toughness, two major points are important to be considered: the effect of a homogeneous microstructure in high-alloy hot work steels, such as ASTM H-series and equivalent steels, and the recent modifications of the H11 with the reduction of silicon, shown, respectively, in Figures 4.9 and 4.10. In terms of homogenizing of the microstructure, it is found in the literature [14,15] that several manufacturing conditions to tool steels may be applied to avoid the presence of undissolved carbides, mainly of V-rich MC type, within the micro-segregated areas as well as intense precipitation of those carbides in grain boundaries, which do not dissolve during hardening (e.g., see Figure 4.9a vs. b). Both effects are caused by the inhomogeneous distribution of alloying elements, as shown in Figure 4.9c, which is caused by the segregation of those elements during solidification. Although segregation is a natural metallurgical phenomenon, the amount of segregation may be reduced by controlling the casting parameters, mainly the casting temperature and cooling rate during solidification (how fast the solidification occurs). For a given the initial segregation, homogenizing treatments may be applied to reduce the segregation effects, as described well in References 14 and 15 (see also the discussion on Figure 3.6). The final result is that the excess amount of V in specific areas is decreased, enabling the dissolution of the coarse carbides prior to forging and resulting in the absence of those carbides in the final microstructure. As seen in Figure 4.9a and b, this leads an increase in toughness by a factor of 3, as shown in Figure 4.17 ahead.

Considering now the second evolution aspect in hot work tool steels, in the last 10–20 years the industry has been using more frequently modified steels with lower amounts of Si and P. In fact, steels H11 and H13 (which evolved from H11) always employed high silicon levels, in the range of 1%, probably by the influence of earlier modifications in 4340 steels where Si was shown to increase the tempering resistance. However, there is an important difference between 4340 steels and H11 (or H13), because in the former the tempering resistance depends on retarding cementite formation while in the latter it depends on the alloy carbides. Therefore, the decrease in silicon content in H11 steels does not necessarily cause a loss in tempering resistance (Figure 4.10a). On the other hand, because of the effect of distribution of alloy carbides, as detailed in References 16 and 17, low-silicon steels

---

* In fact, NADCA 2016 has minimum quality recommendations based on impact toughness of samples heat treated in laboratory conditions and of coupons attached to dies heat treated industrially. For instance, the toughness of a superior H13 is average >13.6 J when laboratory treated but >10.8 J is accepted for industrial coupons, meaning that the quench embrittlement phenomena is accepted up to a given level, but has to be controlled.

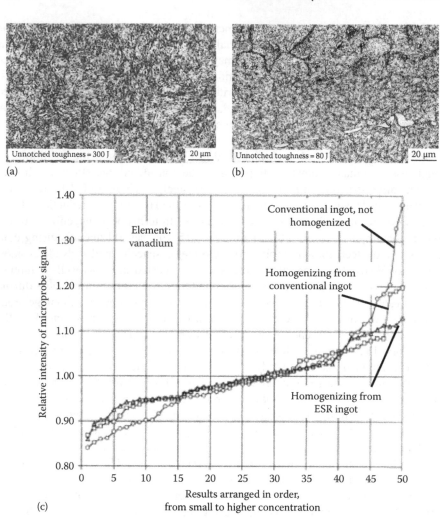

FIGURE 4.9 Microstructure and toughness of a H13 tool steel produced with two different conditions: (a) with ESR remelting plus long soaking times prior to hot forging, to enable better distribution of alloying elements (homogenizing); and (b) conventional casting and direct forging (without homogenizing treatment). The application of homogenizing treatments avoids regions in the microstructure (interdendritic areas) with concentration peaks of allying elements, where undissolved carbides are observed (shown by the arrows in b). (c) Example of this effect of homogenizing treatment in the reduction of areas with high concentration, for the element vanadium. (All results from Mesquita, R.A. and Barbosa, C.A., *Metal. Mater.*, 59, 17, 2003.)

are usually tougher than high-silicon steels. From the microstructure point of view, Cr-secondary carbides change their distribution because of the Si effect in cementite, leading to coarser particles secondary in the martensite interlath areas, which are also found in the fracture surface. The final effect is clearly observed in Figure 4.10b, where for tempering at 600°C (1110°F) the toughness practically doubles when Si is

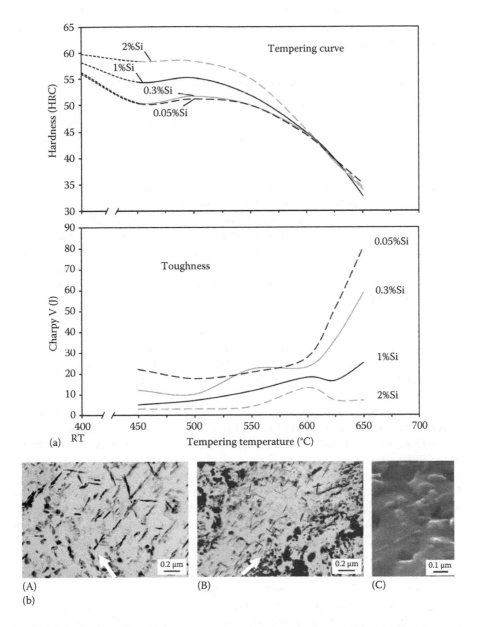

FIGURE 4.10 (a) Effect of Si on the tempering curve and toughness of H11 tool steel. (Adapted from Mesquita, R.A. et al., *Metal. Mater. Trans. A*, 42, 461, 2011.) (b) Distribution of secondary carbides in low silicon (A) and high silicon (B), showing that carbides are concentrated in the martensite boundaries (indicated by the arrows) for the high silicon, which explains the low toughness in those grades, because the same carbides are found in fracture surface (C). (A and B: adapted from Mesquita, R.A. et al., *Metallurg. Mater. Trans. A*, 42, 461, 2011; C: Mesquita, R.A. and Kestenbach, H.J., *Mater. Sci. Eng. A*, 528, 4856, 2011.)

decreased from 1% to less than 0.3%. To maximize toughness, low-Si H11 tool steel also has low P nowadays, which has become common in many applications sensitive to cracking, such as die-casting dies.

4.2.3 WEAR RESISTANCE AND NITRIDING OF HOT WORK TOOL STEELS

It is the opinion of this author that hot strength—also considering tempering resistance—and toughness are the two important properties of tool steels. However, wear resistance is also very important in many applications, especially in hot-forging and extrusion, which involve continuous movement of material on the tool surface, often with the presence of abrasive material such as oxides and scale. It is commonly found in the literature that there is a direct relation between wear resistance and hardness, as compared in Reference 18 for a number of materials.

Therefore, in hot work steels, wear resistance will directly depend on the hardness during operation and also on the combination of initial hardness and tempering resistance. This is in fact the case for quenched and tempered tool steels not containing particles, such as plastic molds and hot work tool steels. In addition, the application of surface treatments, particularly nitriding, is often to used to promote the increase in surface hardness and thus improve the tool life from wear. Because the core hardness is not affected, the interior regions of the tools remain unchanged in terms of toughness. Comparing all the possible surface treatments, nitriding is most commonly applied in hot work tooling, because of the very high developed hardness (above 1000 HV) and because it can be applied in the final tools, involving heating to temperatures of only ~540°C, which is below the common tempering ranges of tool steels, and so not affecting the hardness.

Some examples of developed hardness for different nitriding conditions of H-series tool steels are shown in Figure 4.11a, where typical hardness in the range 1000–1200 HV is reached. The final effect in wear resistance is expected as a result of the direct effect of hardness on this property; but specific data in this sense, namely wear resistance versus nitriding depth, is not common in the literature. Nevertheless, Figure 4.12 presents this correlation, showing the hardness profile for two tool steels and how it is directly correlated with the abrasive wear resistance. One of the steels is a low-alloy tool steel with low Cr, and so the hardness increase by nitriding is not very substantial. The second is a modified tool steel, usually used in plastic molds, but with higher nitriding hardness due to the higher aluminum content and, in turn, also a higher increase in nitrided hardness and wear resistance as a consequence.

In terms of mechanism, the principle is quite simple. If proper conditions for nitrogen dissociation are created at the steel surface, atomic nitrogen can be introduced on the surface and, being a small element, nitrogen may diffuse rather quickly to the interior regions, following the natural tendency to diffuse from a higher to a lower concentration. So, if relatively long times are used during nitriding, which in practice is between 4 and 10 h, the nitrogen introduced into the surface will migrate to the interior regions, but it will be always richer on the surface and continuously decrease toward the center. After a certain depth, no notable differences in nitrogen and its effect on the hardness are found, and this is then considered the nitriding depth. Typically for tool steels, the nitriding depth is between 0.1 and 0.2 mm.

Higher depths are very uncommon and often followed by the embrittlement of the nitrided layers, as shown in Figure 4.12; it is a rather common mistake to target nitriding depths in the range 0.5–1.0 mm, when one is used to carburizing depths. However, it is important to differentiate between the two treatments: carburizing is performed at high temperatures, ~900°C–1000°C (1650°F–1830°F), which enables faster diffusion and thicker layers; the hardening is promoted by the carbon effect on martensite transformation during quenching, and then it is limited to martensite hardness of ~900 HV, which may further decrease after tempering. Nitriding, as mentioned, is performed on already hardened and tempered parts. The strengthening mechanism involves the diffusion and continuous formation of alloy nitrides, which are fine and well distributed, following a mechanism similar to secondary carbides formed during tempering. Because of the high concentration of nitrogen achieved on the surface, hardening due to precipitated nitrides is rather high, achieving easily

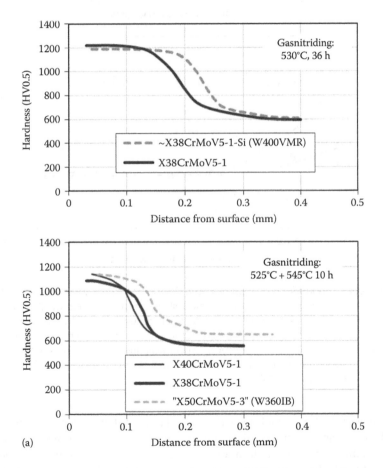

(a)

FIGURE 4.11 (a) Response of different tool steels to nitriding treatments. Correspondence of the steel grades X38CrMoV5-1: H11 mod, X28CrMoV5-1: H11, X40CrMoV5-1: H13. X50rMoV5-3 is a modified H10. (a: From Schneider, R.S.E. and Mesquita, R.A., *Int. Heat Treat. Surf. Eng.*, 5, 94, 2011.)   *(Continued)*

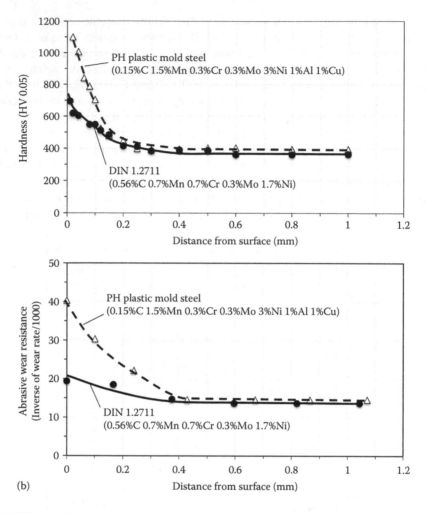

FIGURE 4.11 (*Continued*) (b) Hardness and the corresponding wear resistance of a low-alloy hot work and a precipitation-hardened mold steel, the latter alloyed with Al. Wear test by pin-on-sand paper is the wear resistance considered reciprocal of the wear rate. The measurement of rate is given by the slope of the curve of the worn volume per length of sanding (mm^3/mm), divided by the worn area (mm^2) of the specimen, and so it is dimensionless. (b: From Jarreta, D.D., Report for Internship Training, Program PIEEG, Universidade Federal de São Carlos, São Carlos, Brazil, 2004.)

levels above 1000 HV in tool steels (Figure 4.11). Also, the hardness will depend on the alloying elements forming those nitrides, Cr being of special interest due to its strong bonding to nitrogen to form fine nitrides and due to the often high amount of Cr in tool steels. In certain new developments, especially for extrusion, steels with Al are used, because this alloying element is the strongest in terms of nitride formation and effect on the hardness of tool steels [21]. One example is actually seen in Figure 4.11b, which shows two steels with low amounts of alloying elements but

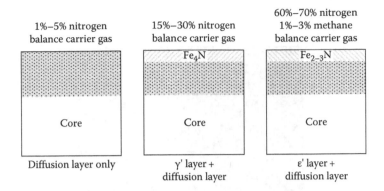

FIGURE 4.12 Microconstituents related to the hardness profile in a nitrided steel. This example is for gas nitriding, but the same reasoning is valid for all nitriding processes. When the potential of nitrogen increases, the microstructure changes from only containing a diffusion layer (represented by the dark points) to the existence of a compound layer (represented by the inclined lines), also named *white layer*, which is very hard but also very brittle. (Adapted from ASM Handbook, *Heat Treating, Section Surface Hardening of Steels*, Vol. 4, ASM International, Handbook Committee, Materials Park, OH, 1991, pp. 387–436.)

one of them with high Al and, as a consequence, with much higher hardness after nitriding.

On the surface, the concentration of nitrogen may achieve such levels that a fine layer, at the very surface, is formed of pure nitrides, which may have different structures depending on the amount of N available. This phenomenon is primarily determined by the method of N introduction at the surface, which may be by a decomposition of a liquid salt based on nitrogen—called *liquid nitriding*—by the decomposition of nitrogen gas—called *gas nitriding*—or by the decomposition of nitrogen molecules by a plasma, called *plasma* or *ion nitriding*. Because of environmental pressures, liquid nitriding is becoming less common, and the two most commonly applied technologies in tool steels are gas and plasma nitriding. In spite of strong differences in terms of methods and equipment, the mechanism and the performance of both may be tuned by the process, in order to achieve the microstructure desired. An example of this tuning is shown in Figure 4.12, where the atmospheric condition was changed to promote a surface layer with or without pure nitrides. This layer is extremely hard but also brittle, and is often not desired in crack-sensitive tool applications, such as die-casting. Because of its ceramic nature, this layer is not etched during metallographic preparation and, when observed under an optical microscope, is very bright, which gives its common name in industry as the "white layer," although the more technical name would be the *compound layer*. Beneath the surface, this gradient of nitrogen concentration, which will lead to a gradient of hardness, is then referred to as the *diffusion layer*.

In terms of actual microstructures, examples are shown in Figure 4.13 (and Figures 4.20 and 4.22), in two conditions of H13 nitrided layer: Figure 4.13a, with only the diffusion layer, and Figure 4.14b with strong formation of a compound layer on the surface but also compounds in the grain boundaries within the

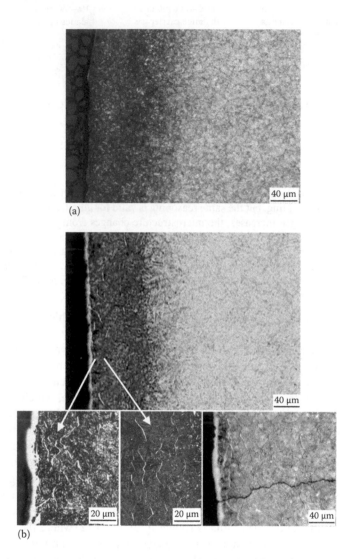

FIGURE 4.13 Example of microstructures in nitride H13 tool steel: (a) without the white layer or precipitates on grain boundaries, and (b) containing the white layer and precipitation of carbonitrides along grain boundaries (see the white phases aligned perpendicular to the growth of nitriding layer). The last image represents another case where the very thick white layer favored the propagation of cracks (to observe the contrast from compounds in grain boundaries, the image was slightly defocused). (All images in (b): Adapted from Canale, L.C.F. et al. (eds.), Failure analysis of heat treated steel components, in: *Failure Analysis in Tool Steels*, Mesquita, R.A. and Barbosa, C.A., eds., American Society for Metals (ASM), Materials Park, OH, 2008, vol. 1, Chapter 11, pp. 311–350.)

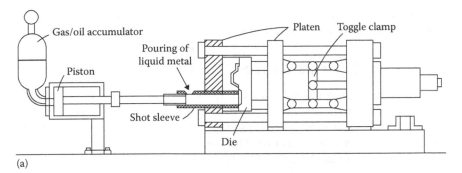

(a)

(b)

FIGURE 4.14 (a) Scheme of a die-casting set, with the common terminology for each part. (b) Example of a die-casting die and a produced aluminum part to be used as an engine component for a vehicle. (a: Modified from ASM Handbook, *Casting*, Vol. 15, ASM International, Handbook Committee, Materials Park, OH, 2008, pp. 724–726; b: From die-casting section of Uddeholm Tool Steel for Die Casting, Available at: http://www.uddeholm.com/files/, Accessed January 2016.)

diffusion zone. This last factor usually leads to a decrease in toughness, because cracks may easily propagate through them, with one picture of cracks shown in detail. Figures 4.20 and 4.22 show specific cases of premature failures, one of the important factors leading to premature cracking being the presence of compounds in the grain boundaries.

Before ending this section, a further comment is warranted on surface treatments. Nitriding was mentioned, in a quite brief way, because of its relevance to hot work tooling applications subjected to wear, such as forging and extrusion. However, there are many other treatments that may be applied to tool steels, which are not within the scope of this book, such as physical vapor deposition (PVD) and others.

4.3 MAIN HOT WORK TOOLING OPERATIONS

Several operations may employ hot work tools, such as forming or casting parts, but three operations are of especial importance because of the volume of tool steel employed: (1) die-casting, (2) hot-forging, and (3) hot-extrusion. In the following sections, these operations are briefly explained, focusing on the parameters affecting the properties and failure mechanisms of tools.

4.3.1 Die-Casting

If one application is to be chosen to represent the challenges imposed on hot work tool steels, die-casting would be the most common choice. This has become even more important currently because the auto industry is eager to reduce weight, and several auto-parts manufacturers are migrating to aluminum castings. Magnesium—as well as zinc (for decorative pieces)—may also be used in some parts, but aluminum is by far the most common alloy used in die-casting, especially for the production of automotive components.

There are two basic types of die-casting technologies: one where the liquid metal and the furnace are integrated into the press, and another where the metal is kept separately and poured into a shot sleeve and then injected into the die. The second option is more common in high-melting-point alloys and thus used commonly in aluminum or magnesium die-casting: one general example is shown in Figure 4.14a. Because of the application of pressure, parts with high complexity may be produced in very short cycles, and the process is very convenient for parts such as gear boxes and engine blocks. One example of a produced part is shown in Figure 4.14b.

There is a close relation between the tool steel used and the die performance, and thus also the process quality and cost, the die's life being dependent on several factors. In fact, die-casting dies and other tools are often subjected to high stresses, leading to several possibilities for failures, which then would require the repair or often the replacement of the die. So, improvement in performance of a tool steel or production process is usually not about avoiding failures, such as in structural steel applications, but extending tool life for postponing a failure, which will still eventually occur. Before going into the details on the die performance and its correlation with tool steel properties, it is important to evaluate the possible failure mechanisms (or end-life mechanisms) that may cause interruption and repair or the complete suspension of the dies under use. These mechanisms may be separated into three main groups in die-casting tools: cracking from thermal fatigue, aluminum adhesion to the tool steel surface, and the erosion of some areas of the tools by the flow of aluminum. These failure modes are, respectively, shown in Figure 4.15a through c and are well described in Reference 24, but are briefly explained and compared with the tool steel characteristics in the following paragraphs.

The die-casting equipment, as shown in Figure 4.14, may involve several components, but most are not necessarily made of tool steel because they are not in contact with the cast metal during the process. On the other hand, the metal parts contacting with the liquid metal are made of tool steel, these parts being highly demanding and often substituted due to the failures, such as the die itself and the shot sleeve. Because of their high cost, sensitivity for failure, and the direct effect on the process productivity and on the quality of the parts, die-casting dies are often the most critical components in terms of failure modes. Specifically in the case of dies, the main cause of failures are cracks from thermal fatigue, shown in detail in Figure 4.15a, and often named *heat checking*, composed of several small surface cracks with no specific propagation paths and also with the appearance of the checking symbol (✓). Because of its importance in die's performance, heat checking will be treated here in more detail than the other failure modes shown in Figure 4.15.

The basic mechanism of heat checking is well described in the literature: it is a typical phenomenon of fatigue cracking, but where the stresses are caused by temperature variations on the surface, and therefore called thermal fatigue. Even when

(a)

FIGURE 4.15 Main failure mechanisms in die-casting. (a) Cracks from thermal fatigue (heat checking), which is the most common failure mechanism for die inserts and cavities. (From Canale, L.C.F. et al. (eds.), Failure analysis of heat treated steel components, in: *Failure Analysis in Tool Steels*, Mesquita, R.A. and Barbosa, C.A., eds., American Society for Metals (ASM), Materials Park, OH, 2008, vol. 1, Chapter 11, pp. 311–350.) *(Continued)*

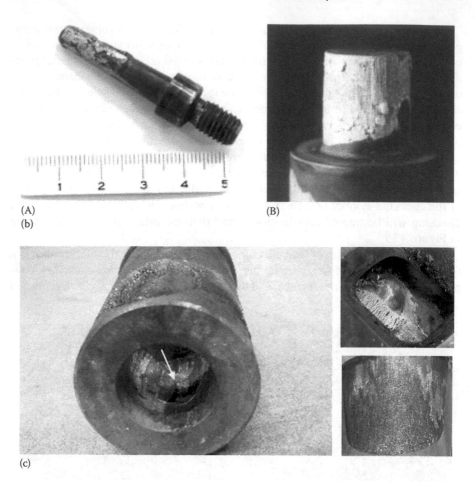

(A)   (B)

(b)

(c)

FIGURE 4.15 (*Continued*) Main failure mechanisms in die-casting. (b) Adhesion of aluminum to steel, also named *soldering*, which is common in small parts such as core ejection. (B: From Uddeholm Tool Steel for Die Casting, Available at: http://www.uddeholm.com/files/, Accessed January 2016.) (c) Erosion and corrosion, caused by the flow of liquid aluminum in high amounts, high speed, and high temperature, a combination that often occurs in the pouring area of shot sleeves. The arrow points the pouring area, where the damage by corrosion and erosion occurs, with specific examples given in the inserts. (From Miglierina, F., Shot sleeve with integral thermal regulation, *Proceedings of the Sixth International Tooling Conference—The Use of Tool Steels: Experience and Research*, Bergstrom, J., Fredriksson, G., Johansson, M., Kotik, O., and Thuvander, F., eds., pp. 1317–1341.)

the dies are preheated to temperatures of, say, 400°C, there will be a substantial temperature gradient when aluminum, at temperatures of about 700°C, reaches the die surface. The layers at the very top surface would tend to increase in dimensions as a result of the natural thermal expansion from 400°C to 700°C, but this expansion is restricted by the layers underneath. The situation is then similar to the application of compressive surface stresses, which "force" the surface layers to have a smaller dimension than would be expected for those temperatures. And, as a consequence, the regions under the surface will be subjected to tensile stresses. After the aluminum solidifies, the surface temperature decreases, but it will be still above the temperature of the interior regions. When the parts are removed, a mold-releasing lubricant is injected, and the surface temperature decreases and tends to be lower than the temperature of the layers underneath, with the thermal stresses then being opposite to that described previously: the surface will be under tension and core under compression. The stresses developed are quite high,* close to or higher than the hot strength of surface layers, and thus cause a typical situation for fatigue: alternation of compressive and tensile stresses (for each produced parts), in a series of usually hundreds of thousands parts produced per die, that is, hundreds of thousands of fatigue cycles.

One way of reducing damages due to thermal fatigue is to reduce the imposed stresses, which in turn means reducing the difference of temperature between the surface and the core. Although this is theoretically possible, in practice a high casting temperature helps filling the small details and then may enable the production of parts with higher complexity. Another alternative to reducing the temperature difference is to employ steels with superior thermal conductivity, such as low-Cr hot work tool steels. Although this alternative exists, other properties are also affected, especially the hardenability and the balance between martensite and bainite. And the variations of thermal conductivity are not so substantial, and choosing low-Cr tool steels for die-casting dies is not common.[†]

Thermal fatigue will then originate at any surface imperfection and will propagate in the dies, which is usually also filled with liquid aluminum at high pressure during casting. This is actually the reason why the cracks are white in Figure 4.15a. The initiation of cracks may also be influenced by microstructural defects, such as undissolved carbides (as shown in Figure 4.9a), which act like micro stress risers and enable the initiation of cracks. After initiation, the propagation of the cracks will depend on the fracture toughness of the steel, because this property is defined and measured by the resistance to the propagation of a preexisting crack. To correlate both phenomena, namely crack initiation and propagation, an impact test is often used. As shown in Figure 4.16, there is a direct correlation between the unnotched impact strength and the resistance to damage of the die. The impact test is actually very important to determine the quality of hot work tool steel, and was the main parameter of the first North American Die-Casting Association (NADCA) recommendation, in 1990 [27]. By not having a notch, the fracture of the specimen is very much dependent on microstructural defects, as explained in Figure 4.9, and the final result will be a proper quantification of the steel quality. It is also important to mention that in Figure 4.16 increasing the hardness tends to increase the resistance

---

* The thermal stress (σ) can be calculated by the expression $\sigma = \alpha \cdot E \cdot \Delta T$, where α is the linear coefficient of expansion of steel, E is Young's modulus, and ΔT is the temperature difference. It is easy to calculate with data on the physical properties of tool steels that the stress will be substantial, of course depending on ΔT, reaching 500–800 MPa and then close to the hot strength of the tool steel (around 800 MPa at a temperature of 600°C, Reference 26).

† However, this may be important in other applications, especially for smaller tools, to reduce damage due to thermal fatigue. See the discussion in Section 4.4.3 on low-Cr hot work tool steels.

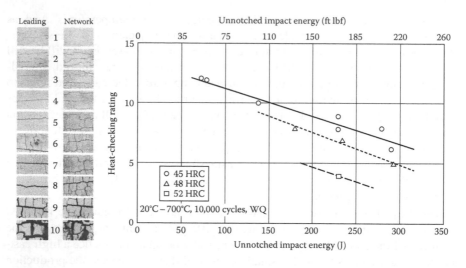

FIGURE 4.16 Effect of both hardness and toughness on the thermal fatigue cracks (heat-checking damage), which is calculated by adding ratings from both columns to determine the final rating of heat checking. Observe that the increase in toughness is directly related to the increase in heat checking resistance, for different levels of hardness. Although higher hardness seems to lead to better results, it is important to remember that when hardness increases, toughness decreases. The best compromise for hardness and toughness and balance of heat checking resistance is usually reached at 45 HRC. (Adapted from Roberts, G.A. and Cary, R.A., *Tool Steels*, 4th edn. Beachwood, OH: American Society for Metals, pp. 645–653, 1980, after Eliasson, L. and Sandberg, O., Effect of different parameter on heat-checking properties of hot-work tool steels, in: *New Materials and Processes for Tooling*, Berns, H., Nordberg, H., and Fleischer, H.-J., eds., Verlag Schurmann & Klagges KG, Bochum, Germany, 1989, pp. 3–14.)

to thermal fatigue, because it increases the strength and thus the resistance to the initiation of fatigue cracks. So, hypothetically, the best situation would be achieving maximum hardness and maximum toughness in tool steels. However, this is not possible because hardness and toughness are inversely related. Therefore, in most casting dies the hardness is chosen at about 45 HRC, which is a compromise between hardness, affecting crack initiation, and toughness, affecting crack propagation.

Because of the improvement in tool steel quality, the use of unnotched impact test has become less common, since the toughness of hot work tool steels can often achieve levels at which the specimen does not break during the test but bend and pass through the machine instead. Therefore, it is today common to measure the toughness of hot work tool steels with the V-notch impact test, the traditional Charpy test, which is used in the new versions of NADCA quality recommendations [11]. The evolution of toughness of H13 and modified H11 steel is shown in Figure 4.17, showing a tremendous improvement in quality by either process (homogenizing or electro-slag remelting, ESR) or compositional changes (Si reduction). A strong improvement would then be expected in the performance today's dies, which in the author's opinion does exist but is often unseen due to other process parameters that currently affect negatively the die's performance,

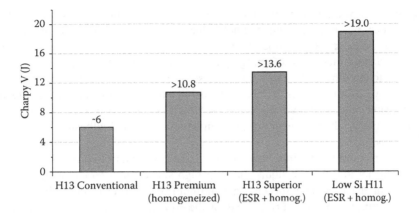

FIGURE 4.17 Evolution of tool steels applied in die-casting. (According to NADCA #229/2006, Special quality die steel & heat treatment acceptance criteria for die-casting dies, North American Die Casting Association—NADCA, Holbrook Wheeling, IL, 2006, 33pp.) The specimens all heat-treated to 45 HRC hardness, with hardening between 1010°C and 1030°C. The conventional H13 was predominant before the 1990s, being the premium level established by NADCA in that year. After 2003, NADCA started with the superior H13 class, and after 2006 with the modified (low-Si, low-P) grades. (Toughness of H13 conventionally would depend on the level of microsegregation, and the values included are from Mesquita, R.A. and Barbosa, C.A., *Metalurg. Materiais*, 59, 17, 2003 (in Portuguese, original title: "Aços ferramenta de alto desempenho para matrizes de fundição sob pressão").)

such as the requirements of higher productivity and more complex parts. Both these factors would lead to an increase in thermal stress, as explained earlier, either by short cycles or by higher casting temperatures, and therefore would tend to decrease the tool life. So, on one hand, tool steels today are much more resistant to die-casting damage but, on the other hand, the process is also more demanding. Often, the tool's life continues to be in the same range of hundreds of thousands of produced parts.

Referring to the other failure modes in die-casting, Figure 4.15b shows an example of small core pins used to produce parts with holes, thereby reducing the number of drilling operations in the cast parts. Because of their small mass and because they are completely covered by aluminum, the core pins are easily heated during one cycle, leading to tempering and loss of hardness. However, a second factor is predominant in core pins, which is the adhesion of aluminum to the pins' surface due to the chemical reaction on the surface between the steel and the cast metal. In the case of aluminum die-casting, it has to be considered that there is a high chemical affinity between iron and aluminum and both elements may form compounds when placed in contact for longer times and high temperatures. This chemical attraction often happens on the surface of core pins, and the practical effect is the observation of strong adhesion of aluminum to the pins' surface,* as shown in Figure 4.15b.

---

* The same effect of adhesion is less common on the die surface because of the higher mass of the dies, which conducts away the heat and avoids strong heating of the surface.

The adhesion of aluminum may cause defects in the produced parts, where imperfections may occur at the produced holes due to breakage of the adhered areas during the parts' extraction, an operation usually done by robots. When the adhesion is strong and combined with softening by the several heating cycles, pins may be bent of even broken during the extraction. The solution lies in avoiding local heating of the pins by design changes or lubricants; using surface treatments of the pins may also be advisable, such as PVD coatings, which create a layer on top of the steel, usually nitrides, thereby avoiding the contact and chemical adhesion between aluminum and iron.

The third important failure mode in die-casting, shown in Figure 4.15c, is caused by the impact and flow of liquid metal at high speed, high temperature, and high amounts at the tool steel surface. In aluminum die-casting of cold chamber, this is very common on the pouring areas of the shot sleeves, as shown by the arrow in Figure 4.15c. In those areas, aluminum in high amounts and just taken from the furnace, at high temperature, is poured, which impacts the steel, causing a mixture of thermal, corrosion (aluminum reacting with iron, as described previously), and washing-out wear, ending up in a damage often called *erosion*, right at the bottom of the poring areas. As shown in Figure 4.14a, after pouring, the piston pushes the liquid aluminum into the die (injection), and if the pouring areas are worn out, it can impair the piston movement or damage it. The final result is that, after several cycles, erosion and corrosion will lead to the need to repair or replace the shot sleeve. The options to improve these failure modes are limited and often involve design changes (reducing the impact of liquid aluminum) or internal cooling, or the use of distinct materials in the pouring areas.

4.3.2 HOT-FORGING

With many specific processes, forging is generally defined as a type of bulk metalworking or a deformation processes in which a metal billet or blank is shaped by tools or dies [29]. A schematic view detailing the blank and the die is shown in Figure 4.18a, and an actual forging die is displayed in Figure 4.18b. In terms of hot-forging, the most common forged materials are carbon and alloyed steels, for example used in the production of auto parts. As shown in Figure 4.18a, several tool steels and also other steels may be used in a die set, the most common being a combination of low-alloy tool steels, such as mod 6F3 (Din 1.2714) for the less demanding parts, and high-alloy tool steels, such as H10, H11, H12, and H13, for the areas subjected to higher temperatures, stresses, and wear. In fact, the forging dies may be subjected to different types of stresses depending on the geometry and the process applied, and this will cause different failure mechanism. In this respect, Figure 4.18c shows a classic scheme of the possible failure modes in a forging die.

The first important mechanism shown in Figure 4.18c is *abrasive wear*, which is caused by the movement of the metal being forged, usually hot steel, on the surface of the die, also containing scale due to the oxidation of the blanks during the heating prior to forging. Being hard and of ceramic nature, the scale acts like a sandpaper on the die surface, scratching and causing material loss by abrasion. As explained in

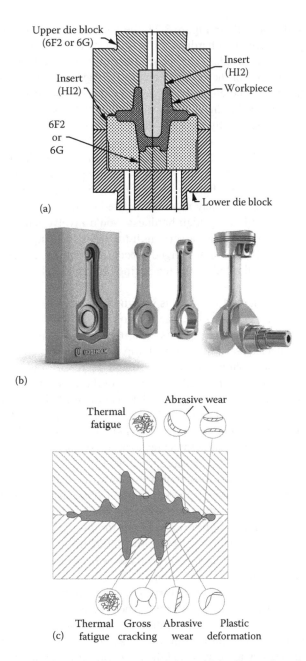

(a)

(b)

(c)

FIGURE 4.18 (a) Example of a die set, containing a high-alloy tool steel (H12) as insert and a low-alloy tool steel (6F2, similar to 6F3 shown in this chapter) as holder and less stressed die. (Modified from ASM Handbook, *Metalworking: Bulk Forming*, Vol. 14A, ASM International, Handbook Committee, Materials Park, OH, 2005, pp. 21–96.) (b) Example of a die made of H13 tool steel, used to produce engine connecting rods. (c) Common failure mechanisms in forging dies. (b, c: From Uddeholm Tool Steel for Forging Applications, Available at: http://www.uddeholm.com/files/, accessed January, 2016.)

Section 3.3.1 and especially in Figure 3.23, the increase in abrasive wear resistance is promoted by an increase in hardness or by the addition of hard particles, particularly undissolved alloy carbides. The second possibility is usually not possible in hot work tool steels because of the strong decrease in toughness and the negative effect on other failure modes. Therefore, the abrasive wear resistance depends solely on the hardness. Increasing the hardness of the die is then the first option to improve the abrasive resistance, and usually hardness in the range of 45–52 HRC is seen in forging dies made of H-series tool steels.

However, hardness beyond 52 HRC is not common, mainly due to the reduction of toughness, as explained in Figure 4.6. It is also important to mention that, even though toughness-dependent failures would allow, hardness increase close to the maximum in hot work steels may not be effective because of tempering during the tool's use. In other words, very high hardness would involve tempering at ~500°C (930°F), which then would cause a very high tendency for tempering and hardness reduction when tools touch the hot-forged piece. In summary, increase in hardness in forging dies is often limited by the tempering effects and also by the decrease in toughness.

One example of failure of this case is shown in Figure 4.19, where hardness was increased to prevent wear, but then cracks also appeared and the increase in wear resistance was also not relevant. In this case, the careful evaluation of the failure showed that the surface areas were being tempered during the operation of the die, resulting in wear. One solution to avoid hardness decrease would be changing the surface heating, but this would lead to changes in the production parameters (e.g., temperatures, cycle time and productivity, lubrication, etc.). The solution chosen in this case was to use a tool steel with higher tempering resistance, allowing the dies to temper to not such high hardness levels (around 50 HRC) but leading to good wear resistance, thus preventing hardness decrease at the surface. An increase in performance of ~50% was observed with this approach [30].

On the other hand, one has to remember that only the surface is worn out by abrasion and so the surface hardness is the key factor. Consequently, methods of surface hardening, very typically nitriding, are very commonly applied in hot-forging dies. The metallurgical parameters of nitriding in hot-forging dies follow the same direction given in the nitriding of tool steels (Section 4.2.3). In case of cracking in nitrided tools, one can avoid too thick layers (keeping below 0.25 mm) and prevent extensive formation of surface white layer and compounds in grain boundaries.

In addition to abrasion, tempering due to a combination of the process heat and high loads may cause *plastic deformation* (permanent deformation), also shown schematically in Figure 4.18c. The solution for this failure mode also lies in the increase in hardness, with similar limitations as pointed out in the previous paragraph: the limitations by tempering and the limitations by toughness. In fact, part of the wear observed in the example of Figure 4.19 was also caused by plastic deformation, and the solution in that case by changing to a steel with better tempering resistance explains that, in terms of plastic deformation, the surface areas would have a more stable strength and then would deform less or deform only after a longer time.

The other mechanisms of failure in tool steels are typically related to cracking, either by *thermal fatigue*, with small and nonoriented cracks, or by *mechanical fatigue*, in which case cracks usually appear in deeper areas where there is a tendency to "open cracks" by tensile stresses. Thermal fatigue cracking in forging dies follows the same mechanism and is affected by the same tool steel properties as explained in Section 4.3.1; that is, to avoid or delay thermal fatigue, a tool steel with high toughness and homogeneous microstructure is desired and the tool should be free from surface stress risers. But, unlike in the case of die-casting, thermal fatigue does not constitute the main cause of failure in forging tools. In many cases, actually, there are failure modes combined with, for example, small surface cracks due

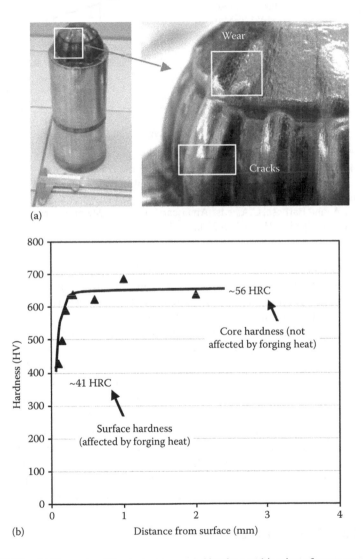

FIGURE 4.19 (a) Example of forging wear caused by the combination of pressure, abrasion, and temperature, leading to (b) hardness decrease in the surface due to heating during forging operations (tempering during forging). *(Continued)*

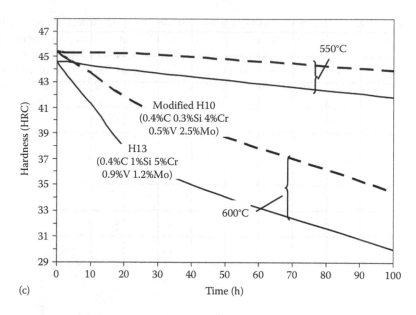

(c)

FIGURE 4.19 (*Continued*)   (c) The improvement in die life was achieved in this case by using a steel with better tempering resistance. (Adapted from Canale, L.C.F. et al. (eds.), *Failure analysis of heat treated steel components*, in: *Failure Analysis in Tool Steels*, Mesquita, R.A. and Barbosa, C.A., eds., American Society for Metals (ASM), Materials Park, OH, 2008, vol. 1, Chapter 11, pp. 311–350.)

to thermal fatigue and larger cracks in areas with stressing conditions, rather than the propagation of one crack. Increasing the toughness, avoiding surface and nitriding defects, and changing the tool design to reduce stresses are typical solutions for failures caused by thermal or mechanical fatigue.

There are also cases with other combinations of failure modes, such as the example in Figure 4.20, with cracks caused by thermal fatigue and abrasive wear. The evaluation of the surface shows that the tools have a very large nitriding layer and that this layer is deeply affected by the heat from forging (Figure 4.20c). In addition, extensive thermal fatigue is also noticed, probably also due to the nitriding conditions and excessive heating. Being a small die, the amount of heat transmitted may be substantial, which would explain the evidence of heating. Therefore, the improvement of the nitrided layer and the decrease in the temperature of the die, for example, by means of a lubricant, are typical solutions for these cases. Using a steel with high tempering resistance may also be advisable, but in this case the solution was already applied and the dies were made of H19, a high-alloy hot work tool steel.

4.3.3   HIGH-TEMPERATURE EXTRUSION

Hot-extrusion is another very important process and area of high use of hot work steels, and is characterized by the production of nonferrous profiles by pushing hot metals through a die orifice (see Figure 4.21a). To increase formability, high temperatures are usually applied, reaching 500°C (930°F) or more in the case of aluminum extrusion. Compared to the processes previously described, namely die-casting and hot-forging, the temperature of the material being formed is much lower, but the contact times, on the other hand, are much higher, being a continuous process. So, mechanical strength at high temperatures is still important in extrusion, although the effects of hardness reduction by tempering during service are less likely.

The container and the auxiliary parts shown in Figure 4.21a are usually made of low-alloy tool steels, typically DIN 1.2714 (mod ASTM 6F3) or similar steels. However, the actual dies, which are in contact with the hot metal (Figure 4.21b and c), are usually made in hot work steels, typically H13, with hardness varying from 45 to 52 HRC. Two main configurations are possible for the dies: the solid type

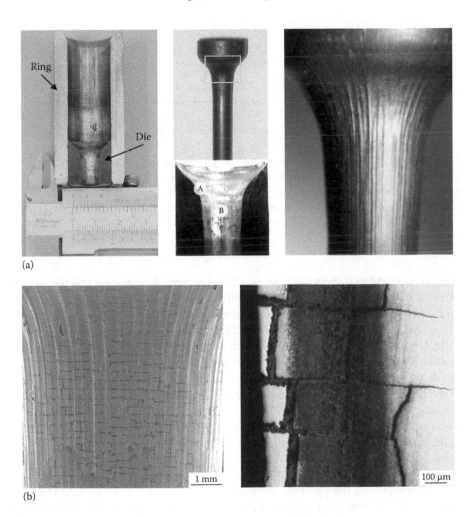

FIGURE 4.20 Example of small forging die after failure, used to produce engine valves. (a) General view of the die, showing the regions with abrasive wear ("A") and thermal cracks ("B"). (b) Detail for the cracks, crossing the nitrided layer. *(Continued)*

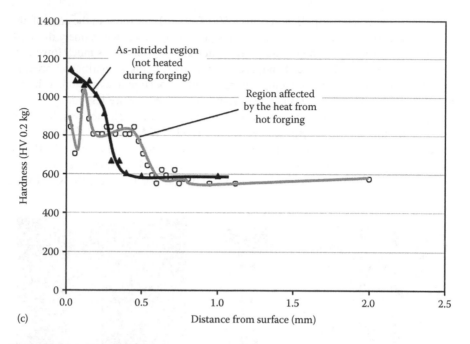

(c)

FIGURE 4.20 (*Continued*) Example of small forging die after failure, used to produce engine valves. (c) Hardness profile in the regions affected by the heat during forging and in the regions not exposed to high temperature. The steel is a modified H13. (Adapted from Canale, L.C.F. et al. (eds.), Failure analysis of heat treated steel components, in: *Failure Analysis in Tool Steels*, Mesquita, R.A. and Barbosa, C.A., eds., American Society for Metals (ASM), Materials Park, OH, 2008, vol. 1, Chapter 11, pp. 311–350.)

(Figure 4.21) and the tubular set (Figure 4.22). The main difference is that the former involves only the passage though a die orifice, while the latter involves a complex movement of the aluminum in the die and reconsolidation into a tube-type profile. The configurations are shown in Figure 4.21.

In terms of performance of the die, the main concern is the wear of the die, mainly the die orifices where the profile is produced, which affects its dimensions. While this is a problem for meeting the specifications, there are also direct effects in the final cost of the produced profiles. In most cases, such profiles are used in the construction industry and are paid for by length and not by weight. This means that when a die wears out, the profile produced is thicker and so heavier; and since the main costly item in profiles is the amount of aluminum used, the production with worn dies adds unnecessary cost. Therefore, avoiding die wear is a critical requirement in extrusion. Hardness could be increased in this sense, but for the same reason as explained in the previous sections, using hardness in the highest range is restricted by the low toughness achieved and also by the possibility of tempering during tool use, though the last effect is relevant not by the high temperatures but by the long heating cycles during the continuous contact of the tool steel with the hot aluminum.

Because of the restrictions in increasing tool hardness, nitriding is extensively applied in extrusion dies, and it would not be an exaggeration to say that practically all aluminum extrusion dies are nitrified. The main reason is the effectiveness of nitriding in increasing the surface hardness, especially in H-type 5%Cr tool steel, as shown in Figure 4.11, which then leads to very significant increase in wear resistance and the die's performance as a consequence. In addition to Cr, new developments have been made using Al as a way to increase nitriding response as well [21].

One important effect of the long exposure times at high temperature in hot extrusion concerns the diffusion of nitrogen in the nitrided layer. As a natural trend in diffusion, elements would move from a region of higher concentration to that of a lower concentration, as long as enough time and activation temperature are provided. Both conditions are met during hot extrusion, so it is common that the layers diffuse

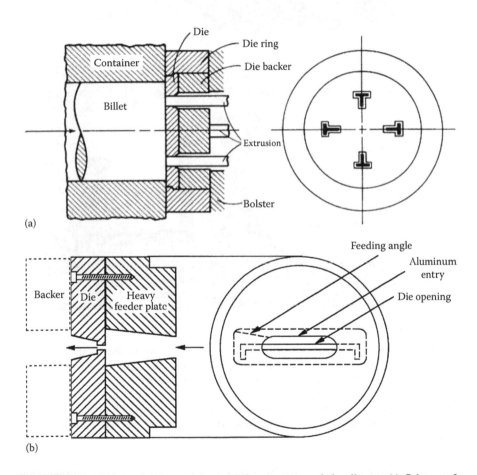

FIGURE 4.21 Schematic view of the extrusion process and the die set. (a) Scheme of a hot extrusion set, showing the aluminum billet pressed inside a container against a die set. (b) Detail for the die set, with the aluminum entry in the feeder plate and the passage through the die. The set may be a two-piece or three-piece set; in the last case it is with a backer after the die (shown in doted lines). *(Continued)*

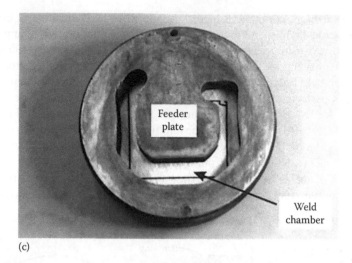

(c)

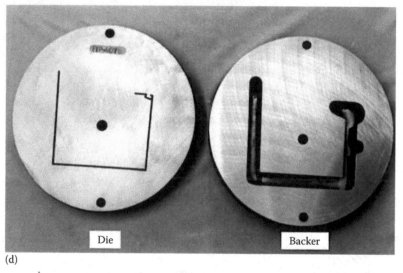

(d)

FIGURE 4.21 (*Continued*)   Schematic view of the extrusion process and the die set. (c) The final die set, with the chamber for the aluminum and the die after it. (d) Opened set, showing the die in detail and the backer that goes after the die, as shown in item a). (a: From Billhardt, C.F. et al., A computer graphics system for CAD/CAM of aluminum extrusion dies, Paper MS78-957, Society for Manufacturing Engineers, 1978; Pictures from b to d: Adapted from Saha, P.K., *Aluminum Extrusion Technology*, ASM International, Materials Park, OH, 2000, pp. 92–105.)

away, leading to a decrease in hardness and decrease of the nitrided hardness, and consequently to a "fading" of its effect in wear resistance. Therefore, it is common that extrusion dies are renitrided after a certain period of use. And for avoiding the diffusion of nitrided layer, the presence of stable nitrides is important, which is important in high-Cr tool steels.

Although wear is the main and natural failure mode in extrusion dies, some specific cases may have failures related to cracks, and then toughness may be important as well. One specific example relates to tubular dies, which are sensitive to cracking at "arms" that hold the mandrel in place. One example of this type of failure is shown in Figure 4.22. In this case, two main factors were related to the lack of toughness: an excess amount of hard compounds in grain boundaries, and excessive hardness due to tempering at low temperatures. These factors lead to the initiation of cracks on the surface (Figure 4.22c) and propagation (Figure 4.22d).

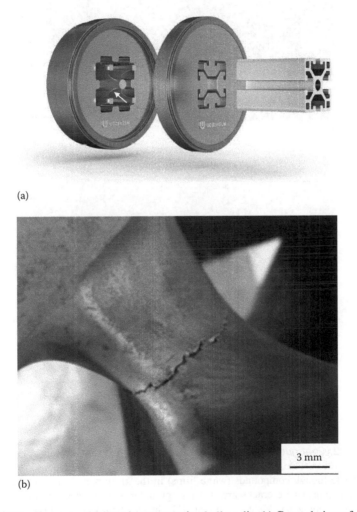

(a)

(b)

FIGURE 4.22 Example of failure in an extrusion hollow die. (a) General view of the extrusion process, where the white arrow indicates the point of failure. (From Uddeholm Tool Steel for Extrusion, Available at: http://www.uddeholm.com/files/, accessed January, 2016.) (b) Photograph of the cracked area. *(Continued)*

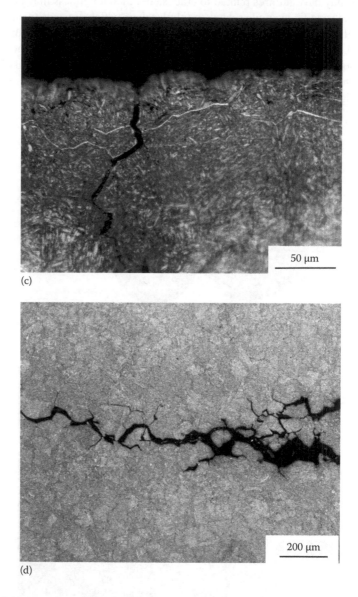

FIGURE 4.22 (*Continued*) Example of failure in an extrusion hollow die. (c) Initial of crack formation at the surface, under an optical microscope, showing that cracks are intergranular and propagate though compounds (white films) in the grain boundaries. (d) Micrograph of an interior region, where cracks are still intergranular and propagating through the prior-austenite grain boundary. Material: low Si H11. Surface hardness: 1100 HV. Core hardness: 51 HRC (about the maximum for this steel). (Courtesy of Villares Metals, Sumaré, Brazil.)

To summarize the performance of the extrusion die, the wear resistance and its association to nitriding are the main factors related to normal wear and failure. On the other hand, excessive loss in toughness or a bad-quality nitriding layer may be associated with unexpected cracks, especially in dies for tubular profiles.

4.4 MAIN GRADES USED IN HOT WORK TOOLING

Considering the properties and requirements previously discussed, it is a common need to decide on the steel for a given application. Therefore, this section describes the main hot work tool steels and the aspects regarding their applications. To initially compare those steels, Figure 4.23 was prepared, showing the three main properties of those steels: (1) hot strength, or more precisely tempering resistance, (2) toughness, and (3) wear resistance. Those steels are discussed in four groups, as follows:

4.4.1 5%Cr H-Series Steels

If one were to select a single steel composition to represent the properties and applications in hot work tooling, this would be the 5%Cr hot work tool steels, remarkably the H13 (DIN 1.2344) and to a lesser extent H11 (DIN 1.2343). With a chemical composition that provides a good balance between cost and properties, these steels are a reference for several applications in hot-forming dies. Originally developed for die-casting, the alloy design of these materials gives high hardenability from relatively low austenitization temperatures (~1020° C), small distortion in heat treatment, minimal tendency to oxidation, good tempering resistance, resistance to erosion by liquid aluminum, and high thermal fatigue resistance. After nitriding, the combination of a homogeneous core hardness of ~45 HRC and surface hardness

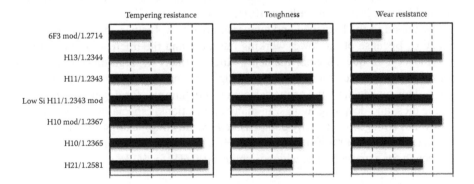

FIGURE 4.23 Qualitative comparison of critical steel properties of the main tool steel (the longer the bar, the better). *Notes:* (1) Tempering resistance is directly related to hot strength and resistance to plastic deformation. (2) Toughness refers to larger dies, this being the reason why 6F3/DIN 1.2714 has the highest values. (3) Wear resistance also considers the nitriding response. (Adapted from Uddeholm Tool Steel for Die Casting, Available at: http://www. uddeholm.com/files/, Accessed January 2016), using the common nitrided hardness of the tool steels for indicating the wear resistance.

>1000 HV (>70 HRC) also bring several advantages in hot wear situations, especially in forging and extrusion.

The difference between H11 and H13 is basically 0.5% more V for the latter, which leads to further improvement in hot hardness and tempering resistance but may also decrease slightly the toughness, especially during quench embrittlement (grain boundary embrittlement after quenching, see Sections 4.2.2 and 3.2.2.2). This greater amount of V in H13 leads to slightly better tempering resistance but slightly lower toughness. And a tendency for inhomogeneous distribution of vanadium carbides may exist in H13 compared to H11, which is of special importance in manufacturing conditions with homogenizing, to maximize the toughness (see Section 4.2.2). That is also the reason for the different levels of quality of an H13 tool steel, as summarized in Figure 4.17, for NADCA limits.

4.4.2 LOW-ALLOYED HOT WORK STEELS

Steels similar to ASTM 6F3/DIN 1.2714 or similar grades are considered low-alloy steels, with typically less than 5% alloying elements. Mainly because of the low Mo and V content, the secondary response hardness and tempering resistance are lower than those of the high-alloyed steels of the H-series. For example, in order to be used, they are usually hot-annealed at ~600°C, and the hardness obtained is between 37 and 40 HRC, while H11 and H13 attain 45 HRC. For some applications, they may be tempered to higher hardness using lower tempering temperatures. But, as already mentioned, this usually does not lead to considerable benefits in terms of hot strength, because of their low resistance to tempering. This is the main reason that restricts the application of low-alloy tool steels in die-casting dies.

On the other hand, the low alloy content of those steels makes them economically interesting, especially for large tools, mold, and die holders or bases. Because the working hardness is usually of 40 HRC or less, low-alloy hot work steels are often supplied in the hardened and tempered condition, further contributing to cost reduction. Because of the lower working hardness and the microstructure with the low tendency for quench embrittlement (which is likely in large dies), DIN 1.2714 and modified steels tend to present high toughness in large hardened and tempered blocks. This makes them the primary steel used in hammer forging dies, other large forging dies, and deep engravings. In many applications, especially in forging dies, the low hot strength is partially compensated by nitriding, leading to surface hardness in the range of 600–700 HV (~55–60 HRC). However, this nitrided hardness is not enough for extrusion dies, which is about 1100 HV for H13.

4.4.3 LOW-Cr HOT WORK STEELS: H10 AND MODIFIED GRADES

As mentioned in Section 4.3.1, the resistance of tool steels to thermal fatigue and thermal shock depends mainly on three factors: the hot strength (and tempering resistance) of the surface layers, toughness, and thermal conductivity. The hardness and toughness above presented exist in 5%Cr steel AISI series H. However, the high Cr content leads to a decrease in thermal conductivity for those grades. For example, the thermal conductivity for H11 or H13, with 5%Cr, is of about 26 W/m·K, while for H10, with 3%Cr, it is ~32 W/m·K [35]. Thus, ASTM H10, DIN 1.2365, DIN 1.2885 (similar to 1.2365, but with Co), or the recently developed grades are commonly used in applications where high thermal conductivity is necessary. The reduction in chromium content would also offer a better potential of precipitation hardening but, as mentioned previously, negatively affects the hardenability and the amount of martensite in comparison to bainite [35]. In some of those grades, the high Mo content would tend to overcome this effect in hardenability but not fully because of the strong reduction in Cr, from 5% to usually less than 3% in low-Cr, high-alloy tool steels. So, the use of low-Cr hot work steel is still limited.

Given the benefits of thermal shock resistance and hot strength,* the H10 modified low-Cr hot work tool steels are normally used in dies employed at high temperatures and high exposure to thermal shock, especially in small tools (<100 mm or 4 in. thickness) that are highly water cooled. A typical example is their use in high-speed presses (known as Hatebur). Another application that is showing increased importance today is tools for press-hardening (hot-forming) of automotive parts such as B-pillars and doors reinforcements. In those applications, as in any hot-forming process, the metal sheet is heated and formed at high temperature inside a die; the difference is that the steels used in press-hardening are usually heat-treatable MnB-alloyed steels. In this sense, after forming, the parts are kept inside the dies, leading to fast cooling to room temperature. Although the cooling speeds are not as high as in water or oil quenching, the dies are water cooled and they promote sufficient quenching to promote martensite formation and thus the formed parts will show a considerable increase in strength in the final geometry. Therefore, the tool steel used, in addition to withstanding the pressures and wear, should have high thermal conductivity for heat extraction.

4.4.4 HIGH-MO OR HIGH-W TOOL STEELS

As explained in Section 4.2, the hot strength is properly represented by the tempering resistance, which in turn depends on the precipitation hardening (see Section 3.2.2.3). As shown in Figure 3.20, the elements Mo, V, and W are the main formers of fine and stable precipitates and so in strengthening by precipitation hardening. V has a limit of solubility of ~0.5%, for a carbon content of ~0.4%. So, the amount of V for this purpose is limited. Therefore, several tool steels present high amount of either Mo, or W, or both, in addition to 0.5%V, to improve precipitation hardening. Besides, to dissolve all the carbides for subsequent precipitation during tempering, the hardening temperatures may need to be increased in high-alloy hot work steels. Examples of this combination of chemical composition and hardening conditions

---

* Though not expected, the decrease in Cr also increases the tempering resistance, as shown in some literature [N12, N13]. In the author's opinion, the mechanism is not totally clear, but it may be affirmed that it is based on the (negative) Cr effect in the precipitation hardening led by Mo and V. Some reasons are (1) the dissolution of Mo and V in Cr-rich carbides, which are bigger and less important to secondary hardening, but drag the Mo content [36], or (2) the effect of Cr when in solution in M_2C carbides and changes in the lattice parameter of this carbide, which according to Reference 37 leads to weaker precipitation hardening.

can be found in steels such as AISI H19, H20, H21, H42, and other very high alloy tool steels for hot work tooling.

Given the high hot wear resistance, these tool steel grades are typically employed in forging alloy steel or extrusion where higher temperatures or long exposition times are involved. However, their relatively low toughness and high cost limit the widespread use of high-W steels. The low toughness, especially in larger dies, is a result of the strong tendency in these grades for quench embrittlement (precipitation on grain boundaries), as explained in Sections 4.2.2 and 3.2.2.2. The high cost is directly associated with the amount of Mo, W, or V, which are the most expensive alloying elements. In addition, those grades have lower Cr, which is not a problem for hardenability due to the high content of other elements and improves even more the tempering resistance; but the low Cr content may compromise the hardness after nitriding, which tends to be lower than for 5%Cr steels.

The present chapter ends here, after explaining the main properties of hot work tool steels, the most common applications, and the failure mechanisms and their correlation with the properties and heat treatment or nitriding conditions. At the end, a brief discussion of all those grades was given, for the 5%Cr H-series, the low-alloy, low-Cr steels, and the highly alloyed tool steels for hot work tooling.

REFERENCES

1. L.C.F. Canale, R.A. Mesquita, G.E. Totten (Eds.). Failure analysis of heat treated steel components. *Failure Analysis in Tool Steels*, Eds. R.A. Mesquita, C.A. Barbosa, Materials Park, OH: American Society for Metals (ASM), 2008, vol. 1. Chapter 11, pp. 311–350.
2. ÖNORM EN ISO 4957:1999. *Tool Steels*. Publisher and printing: Österreichisches Normungsinstitut, 1020 Wien.
3. ASTM A600-92 (reapproved 2010). Standard specification for tool steel, carbon. Materials Park, OH: ASTM International.
4. C.F. Jatczak. Specialty carburizing steels for high temperature service. *Metal Progress*, 113, 70–78, April 1978.
5. G.A. Roberts, R.A. Cary. *Tool Steels*, 4th edn. Beachwood, OH: American Society for Metals, 1980, pp. 645–653.
6. Uddeholm Tooling. Technical catalogue for tool steels. Available online at: http://www.uddeholm.com/files/. Accessed December 2015.
7. J.H. Hollomon, L.D. Jaffe. Time-temperature relations in tempering steel. *Transactions of the Metallurgical Society of AIME*, 162, 223–249, 1945.
8. Böhler Edelstahl. Technical catalogue for tool steels. Available online at: http://www.bohler-edelstahl.com/english/1853_ENG_HTML.php. Accessed December 2015.
9. Villares Metals. Technical catalogue for tool steels. Available online at: http://www.villaresmetals.com.br/pt/Produtos/Acos-Ferramenta. Accessed December 2015.
10. J.C. Hamaker, Jr. Die steel useful for ultra high-strength structural requirements. *Metal Progress*, vol. 70, p. 93, December 1956.
11. NADCA #229/2006. Special quality die steel & heat treatment acceptance criteria for die casting dies. Holbrook Wheeling, IL: North American Die Casting Association—NADCA, 2006, 33pp.
12. M.L. Schmidt. Effect of austenitizing temperature on laboratory treated and large section sizes of H-13 tool steels. *Proceedings of Tool Materials for Molds and Dies*, Eds. G. Krauss, H. Nordberg, Golden, CO: Colorado School of Mines Press, 1987, pp. 118–164.

13. R.A. Mesquita, C.A. Barbosa. High performance steels for pressure die-casting dies. *Metalurgia & Materiais*, 59, 17–22, 2003 (in Portuguese, original tittle: "Aços ferramenta de alto desempenho para matrizes de fundição sob pressão").

14. S. Wilmes, K.P. Burns. Comparison of the toughness of hot work tool steel from different manufacturing processes with regard to the use in die casting dies.). *Gießerei*, 76(24), 835–842, 1989. (Original in German, title: Vergleich der Zähigkeit von Warmarbeitsstahl unterschiedlicher Herstellverfahren im Hinblick auf die Verwendung für Druckgießformen).

15. H. Berns, E. Harberling, F. Wendl. Influence of the annealed microstructure on the toughness of hot work tool steels. *Thyssen Edelstahl Technische Berichte*, 16, 45–52, May 1990.

16. R.A. Mesquita, C.A. Barbosa, E.V. Morales, H.J. Kestenbach. Effect of silicon on carbide precipitation after tempering of H11 hot work steels. *Metallurgical and Materials Transactions A*, 42, 461–472, 2011.

17. R.A. Mesquita, H.J. Kestenbach. On the effect of silicon on toughness in recent high quality hot work steels. *Materials Science & Engineering A*, 528, 4856–4859, 2011.

18. M.M. Khrushchov, M.A. Babichev. An investigation of the wear of metals and alloys by rubbing on an adhesive surface. *Friction and Wear in Machinery*, 12, 1–13, 1958.

19. R.S.E. Schneider, R.A. Mesquita. Advances in tool steels and their heat treatment, Part 2: Hot work tool steels and plastic mould steels. *International Heat Treatment and Surface Engineering*, 5, 94–100, 2011.

20. D.D. Jarreta. Report for internship training, Program PIEEG. São Carlos, Brazil: Universidade Federal de São Carlos, 2004.

21. J.B. Bacalhau, C.A. Barbosa. Characteristics of the new developed hot work tool steel for aluminium extrusion. *Proceedings of the Ninth International Conference on Tooling: "Developing the World of Tooling"*, Loeben, Austria, 2012, pp. 175–182.

22. ASM Handbook. *Heat Treating, Section Surface Hardening of Steels*, vol. 4. Materials Park, OH: ASM International, Handbook Committee, 1991, pp. 387–436.

23. ASM Handbook. *Casting*, vol. 15. Materials Park, OH: ASM International, Handbook Committee, 2008, pp. 724–726.

24. Uddeholm Tool Steel for Die Casting. Available at: http://www.uddeholm.com/files/. Accessed January 2016.

25. F. Miglierina. Shot sleeve with integral thermal regulation. *Proceedings of the Sixth International Tooling Conference—The Use of Tool Steels: Experience and Research*, Eds. J. Bergstrom, G. Fredriksson, M. Johansson, O. Kotik, F. Thuvander, Karlstad, Sweden, 2002, pp. 1317–1341.

26. R. Sivpuri. *Dies and Die Materials for Hot Forging. ASM Handbook*, vol. 14A, Metalworking: Bulk Forming, Materials Park, OH: ASM International, 2005, pp. 47–52.

27. NADCA 207/1990. Premium quality H13 steel acceptance criteria for pressure die-casting dies. River Grove, IL: North American Die Casting Association—NADCA, 1990, 12pp.

28. L. Eliasson, O. Sandberg. Effect of different parameter on heat-checking properties of hot-work tool steels. *New Materials and Processes for Tooling*, Eds. H. Berns, H. Nordberg, H.-J. Fleischer, Bochum, Germany: Verlag Schurmann & Klagges KG, 1989, pp. 3–14.

29. S.L. Semiatin. Introduction to bulk-forming processes, ASM Handbook, vol. 14A, *Metalworking: Bulk Forming*, Materials Park, OH: ASM International, 2005, pp. 1–7.

30. R.A. Mesquita, L.C. França, C.A. Barbosa. Cases of new hot work tool steels. *Tecnologia em Metalurgia e Materiais*, 2, 70–75, 2005 (in Portuguese).

31. ASM Handbook. *Metalworking: Bulk Forming*, vol. 14A. Materials Park, OH: ASM International, Handbook Committee, 2005, pp. 21–96.

32. Uddeholm Tool Steel for Forging Applications. Available at: http://www.uddeholm. com/files/. Accessed January 2016.
33. C.F. Billhardt, V. Nagpal, T. Altan. A computer graphics system for CAD/CAM of aluminum extrusion dies. Paper MS78-957, Dearborn, MI: Society for Manufacturing Engineers, 1978.
34. P.K. Saha. *Aluminum Extrusion Technology.* Materials Park, OH: ASM International, 2000, pp. 92–105.
35. I. Schruff. *Comparison of Properties and Characteristics of Hot-Work Tool Steels X 38 CrMoV 5 1 (Thyrotherm 2343), X 40 CrMoV 5 1 (Thyrotherm 2344), X 32 CrMoV 3 (Thyrotherm 2365) and X 38 CrMoV 5 3 (Thyrotherm 2367).* Thyssen Edelstahl Technische Berichte, Special Issue, pp. 32–44, May 1990.
36. R.A. Mesquita, C.A. Barbosa. Tool steel for extrusion dies. Patent US 2013/0243639, Priority: 2010.
37. F.B. Pickering. *Physical Metallurgy and the Design of Steels.* London, U.K.: Applied Science Publishers Ltd., 1978, pp. 136–138.
38. Uddeholm tool steel for extrusion. Available at: http://www.uddeholm.com/files/. Accessed January 2016.

5 Cold Work Tool Steels

5.1 INTRODUCTION TO COLD WORK TOOLING

Cold work tooling is often defined as any forming operation in which the formed material is at temperatures below 200°C (390°F), more often at room temperature. Several forming processes are typical in cold work tooling, such as metal shaping by stamping, coining, shearing, and blanking (Figure 5.1). Especially remarkable are the cold work tooling processes of steel sheets and other formats, mainly low-carbon steels designed to provide extended ductility and thus the ability to be cold-worked to a large extent.

This limit of 200°C (390°F) exists because, in practical terms, the temperature will not affect the process and the conditions of the workpiece, and therefore will not lead to variations on tool properties, tool surface, and the surface of the formed piece. So, in comparison to hot work tooling, there are several advantages related to the quality of the produced part, in terms of a shiny surface appearance and also very precise dimensions. On the other hand, the forces are much higher and the amount of deformation that can be applied to the formed part is limited in cold work, when compared to hot work tooling.

The tool steels used in cold work are of several different AISI classes, and a variety of grades exist.* However, in Table 5.1, the main grades are shown with their compositions. Two important factors are clear: The high carbon content of ~0.8%–1.0%, which later in this chapter will be shown to result in high hardness; and the lower amount of precipitation hardening elements, such as V and Mo, because those grades are usually not designed to generate secondary hardness during tempering and no resistance to tempering when used at low temperatures.

5.2 FAILURE MECHANISMS AND BASIC PROPERTIES OF COLD WORK TOOL STEELS

The fact that forming operations are performed not at high temperatures simplifies the understanding of the tool steel behavior, because the properties are hardly ever affected by the application of the tool (disregarding some specific effects of retained austenite). Therefore, the different operations are not described individually, as was done for the hot work tool steels (Chapter 4), but are done in terms of the mechanisms that lead to the end life of a tool, as discussed in this section.

There are several causes of damage to tools and end-life mechanisms on cold work tools, but most may be classified as due to a single possibility or a combination

---

* There are actually other materials that can be used in cold-forming tools, such as cast irons, hard metals (solid carbides), or, in very specific cases, ceramic materials. Because of the nature of this book, this chapter will only deal with tool steels used in cold work tooling.

(a)

Hub cover for truck

Tool for deep drawing

(b)

Progressive tool

FIGURE 5.1 Examples of cold work tools: (a) deep drawing tool set and example of produced part. (b) A progressive tool, for high-speed production of several forming operations (e.g., punching, coining, bending or cutting), combined with an automatic feeding system. (From Uddeholm Tool Steel for Cold Work Tooling, Available at: http://www.uddeholm.com/files/, Accessed January 2016.)

TABLE 5.1
Chemical Composition of the Main Cold Work Tool Steels Applied in Industry, Showing the Commonly Observed Target Values

Hot Work Tool Steels			Most Common Chemical Composition (wt.%)							
ASTM	EN/DIN	UNS	C	Si	Mn	Cr	V	W	Mo	Other
O1	1.2510	T31501	0.9	0.3	1.2	0.5	0.1	0.5	—	—
W1	1.1545	T72301	1.0	0.2	0.2	—	—	—	—	—
W2	1.1645	T72302	1.0	0.2	0.2	—	0.25	—	—	—
A2	1.2363	T30102	1.0	0.3	0.7	5.0	0.25	—	1.1	—
D2	1.2379	T30402	1.5	0.3	0.3	12.0	0.9	—	0.9	—
D3	1.2080	T30403	2.1	0.3	0.3	12.0	—	—	—	—
D6	1.2436	—	2.1	0.3	0.3	12.0	0.1	0.7	—	—
8%Cr (not standardized)			0.9	0.9	0.5	8.0	0.5	—	2.2	+Nb or Al
S1	1.2550	T41901	0.5	0.8	0.2	1.3	0.2	2.0	—	—
S7	—	T41907	0.5	0.3	0.5	3.2	—	—	1.5	—

Note: The specific limits and ranges are given by the standards ASTM A681, A and Euro Norm ISO 4957 (old DIN 17350 and EN 10027), according to References 2 and 3. Throughout this chapter, the DIN number will be referred to.

of possibilities, as shown in Figure 5.2. The main cause is wear, which may be defined in many different ways, but in this book it is simply treated as a loss of dimensions of the tool due to the partial removal of material on its surface. Two main mechanisms are identified in cold work tooling wear (Figure 5.2a): abrasive or adhesive. The main difference is that, during abrasion, a nonmetallic particle scratches the tool surface, leading to grooves and material loss. And, in the second, adhesion of the workpiece to the tool surface may lead to different effects, with tool damage and also adhesion of the formed material on the tool surface (galling). The second failure mode, shown in Figure 5.2b, is related to cracks, with major cracks that cause immediate interruption of the tool's use, or microcracks, which lead to loss of tool dimension, especially in sharp edges. And the last—less common—is the deformation of the tool's working areas (Figure 5.2c), especially in sharp regions, caused by high forces concentrated in a small area, mainly related to the high forces necessary to deform parts at room temperature.

All these mechanisms are carefully discussed in this section, with especial regard to the tool steel properties and forming conditions related to a specific failure. It may be seen that, even being sometimes a combination of several mechanisms, the understanding of the end life gives important inputs to understand the needed change in the tooling operation and, of especial concern in this book, on the tool steel, such as increasing the hardness, improving the carbide distribution, decreasing the hardness, increasing the toughness, and so on.

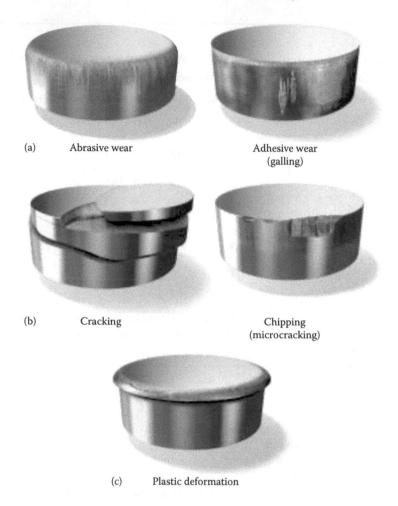

(a)   Abrasive wear   Adhesive wear
                                                  (galling)

(b)   Cracking   Chipping
                                               (microcracking)

(c)   Plastic deformation

FIGURE 5.2 Examples of the main failure mechanisms for cold work tool steels: (a) material loss by wear, which can be either abrasive or adhesive wear. (b) The propagation of cracks, with extensive cracking leading to catastrophic failure or microcracks/chipping, on the sharp ends. (c) Plastic deformation, caused by the high stresses developed during the forming process. (Modified from Uddeholm Tool Steel for Cold Work Tooling, Available at: http://www.uddeholm.com/files/, Accessed January 2016.)

5.2.1 STRENGTH AND HARDNESS

One important characteristic of cold work tool steels is the achievement and the application of high hardness. While this is correct, a full understanding on the need for high hardness is important. As will be shown later, this property is directly related to both abrasive and adhesive wear, the most common failure mode in cold work tooling, and also affects negatively the toughness and so the likelihood of cracking, the second most common failure possibility. However, the third mechanism, plastic deformation, is discussed here as the basis of the need for high hardness, and then

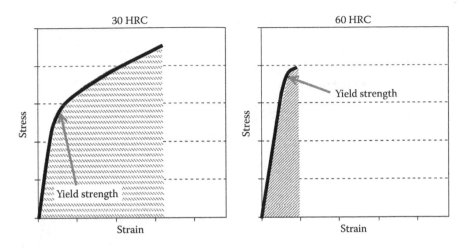

FIGURE 5.3 Schematic stress–strain diagram for cold work tool steels. Observe that the increase in hardness has a direct effect on the increase of yield strength but an inverse effect on the area under the curve, which represents the absorbed energy and so the toughness.

its implications to the other mechanisms (wear or cracking) are addressed (Sections 5.2.2 and 5.2.3).

Figure 5.3 shows a schematic representation of the effects of hardness increase in a generic stress–strain diagram, pointing that when hardness increases, the yield strength also increases. Therefore, if a given cold work tool is showing areas with dimensional changes during work—typically sharp corners being rounded—this is a clear indication of plastic deformation and so of insufficient hardness, disregarding the possibilities of changes in tool design or forming forces. In other words, the stresses on those areas exceed the yield point of the tool steel, in a way that the material is brought to the plastic regime and starts plastic (permanent) deformation. And the increase in hardness, such as shown with the shift from 30 to 60 HRC in Figure 5.3, will require more force (stress) to initiate plastic deformation in a given tool steel, thus preventing possible failures by plastic deformation. However, as explained in Section 5.2.3, this will lead to a decrease in toughness (represented by the area under the curve) and consequently higher sensitivity for cracking.

Returning to the discussion on plastic deformation, the presence of high tooling forces in cold work tooling, due to the high strength of the formed material (which is usually at room temperature), leads to a need for hardness usually above 60 HRC for cold work tool steels. This is a condition common for many grades and, being important only at low temperatures, may be easily achieved by high-carbon grades after quenching and tempering at low temperatures. In this respect, Figure 5.4 shows clearly that several cold work steels, when tempered at low temperatures, may achieve 60 HRC or more, which is primarily determined by the martensite hardness, which in turn depends on the carbon content: for C > 0.7%, those hardness levels will be easily achieved (shown later in Figure 5.10). So, by understanding this reasoning, the correlation between high tooling forces, high hardness, high carbon

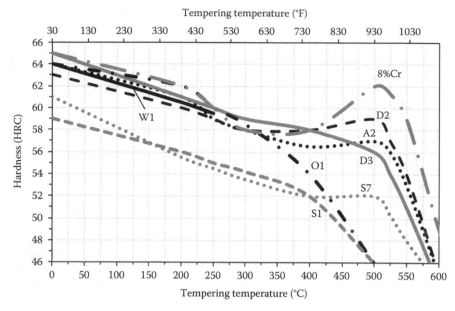

FIGURE 5.4 Tempering diagrams for the most common cold work tool steels, showing hardness after two times tempering for 2 h. Data from brochures of traditional tool steel producers. (From Uddeholm Tooling, Technical catalogue for tool steels, Available online at: http://www.uddeholm.com/files/, Accessed December 2015; Böhler Edelstahl, Technical catalogue for tool steels, Available online at: http://www.bohler-edelstahl.com/english/1853_ ENG_HTML.php, Accessed December 2015; Villares Metals, Technical catalogue for tool steels, Available online at: http://www.villaresmetals.com.br/pt/Produtos/Acos-Ferramenta, Accessed December 2015.)

content, and low tempering temperature becomes clear, and therefore corresponding to their application only in cold work tooling.

With the forming temperatures close to room temperatures, the only part of the tempering chart that would apparently matter would be the tempering at or below 200°C (390°F). While this is partially true, and most grades show a dramatic decrease in hardness after tempering above 400°C (750°F), new grades such as the 8%Cr steels have been developed to enable tempering above 500°C (932°F) (Figure 5.4). There are two main reasons for this change in tempering strategy:

1. When the tool steel is tempered at high temperatures, the hardness is (less) dependent on the martensite transformation but also on the precipitation of fine carbides. Martensite being a very hard but very brittle phase, the relief of martensite microstructure by tempering at high temperatures leads to a substantial increase in toughness (Figure 5.8).
2. As explained in Section 4.2.3, the increase in surface hardness by surface treatments such as nitriding or physical vapor deposition (PVD) may lead to important benefits in wear resistance, without compromising the overall toughness. However, those treatments are usually applied at temperatures above 500°C (930°F), which would lead to a substantial decrease in hardness in most cold work steels (Figure 5.4). The strong precipitation hardening of 8%Cr steel overcomes this problem and makes those grades ideal for application in surface treatments or coatings.

With high- or low-temperature tempering, most cold work steels are tempered to very high hardness, as explained here, to improve resistance against plastic deformation, but which also affects the wear resistance (next section). And this characteristic of extremely high levels of hardness also explains the application of other steels, such as high-speed steels, in cold work tooling. With the same advantage of high-temperature tempering on surface treatment as the 8%Cr steels, high-speed steels are even harder, and achieve commonly 64 HRC or more. The need for extreme high levels of hardness also leads to the application of other materials, in very specific applications in cold work tooling, such as hard metals (composites of carbides in a metallic matrix) or even ceramic materials. High-speed steels are not discussed here but in Chapter 6; however, it may be motioned that they follow the same strategy to promote a high-hardness matrix, with carbides distributed, with the difference of necessarily promoting high-temperature strength (hot hardness).

5.2.2 Wear Resistance

5.2.2.1 Abrasive Wear

Abrasive wear is defined by ASTM [7] as hard particles or hard protuberances that are forced against and move along a solid surface, causing damage and progressive material loss to the solid surface. In tooling, abrasive wear is often found when a nonmetallic particle, such as an oxide or ceramic grain, scratches the surface of the tool. This is due the presence of impurities such as dust or oxidation in metal-forming operations, or by the tooling operation itself when ceramic or other nonmetallic materials are shaped in a tool-steel die. Although pure abrasive conditions may exist, such as in the last example, most cases of cold-work tooling involve at least partial abrasive wear, to a degree that usually depends on the operation and lubrication conditions.

In terms of a microstructural approach, the wear may be understood by the movement of an abrasive grain (abrasive agent) on top of the tool surface, as exemplified earlier in Section 3.3.1, and this approach is again used here. In abrasive conditions, comparing several metallic materials, the wear is directly related to the material hardness, as shown in Figure 5.5a. However, the increase in hardness in tool steel is limited by the maximum achieved in ferrous alloys, which is ~65–70 HRC (900–950 HV) by a high-carbon martensite with precipitates. Therefore, hard particles—as hard as or even harder than the abrasive agent—are often added to the steel microstructure. This may be done by powder mixtures, but the most common way is the *in situ* formation of carbides within the tool steel microstructure, by the combination of carbon and alloy elements during the solidification (it is strongly recommended to refer to Section 3.3, for all the explanation of these mechanisms). In addition to the hardness, other factors also affect the effect of particles in a steel microstructure, as depicted in Figure 5.5b. In tool steels, the net effect is then given by the presence of hard carbides, which are held by a high-hardness microstructure, and the combined effects of high hardness and hard particles, as shown in Figure 5.5c.

Of all the parameters presented, the most important are related to the hardness as well as the size and distribution of the undissolved carbides. The details of these are specifically addressed in Section 5.3, but here just the wear effect is considered. In terms of hardness, cold work tool steels present two main types of hard and undissolved carbides: Cr-rich M_7C_3, with about 1700 HV, and V- or Nb-rich MC types,

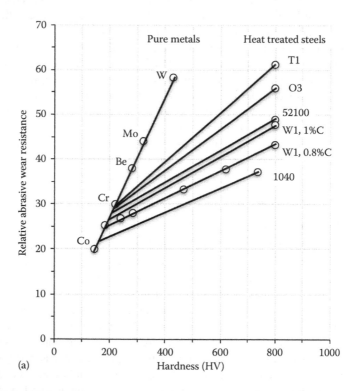

(a)

FIGURE 5.5 Main microstructure parameters affecting the abrasive wear resistance. (a) Matrix hardness, showing that for a wide range of materials the wear resistance increases when hardness increases. (Built with data from Khrushchov, M.M. and Babichev, M.A., An investigation of the wear of metals and alloys by rubbing on adhesive surface. Friction and Wear in Machinery, American Society of Mechanical Engineers, ASME, Vol. 12, pp. 1–12, 1956.)   (*Continued*)

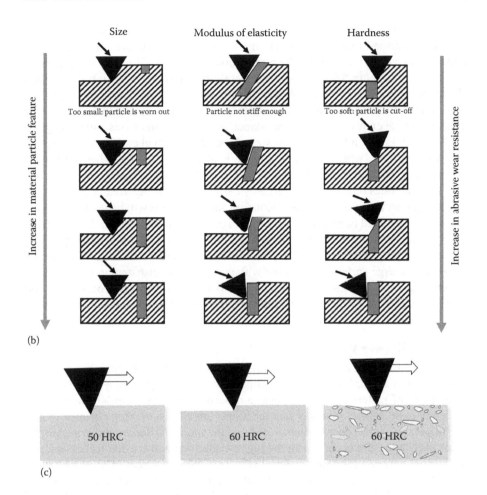

FIGURE 5.5 (*Continued*)   Main microstructure parameters affecting the abrasive wear resistance. (b) Schematic view of a hard abrasive agent encountering particles present in the material surface, with different particle properties. (Modified from Zum Gahr, K.-H., *Microstructure and Wear of Materials*, Elsevier, New York, 1987.) (c) Combination of both, showing that the ideal situation is a hard microstructure matrix and the presence of well-distributed, large particles (Modified from Canale, L.C.F. et al., Failure analysis of heat treated steel components, in: *Failure Analysis in Tool Steels*, Mesquita, R.A. and Barbosa, C.A., eds., American Society for Metals (ASM), Materials Park, OH, Vol. 1, Chapter 11, pp. 311–350, 2008) and explained earlier in Figure 3.23.

with up to ~2400 HV. V-rich MC carbides usually dissolve other elements, which leads to a decrease of this hardness to ~2000 HV [8–10]; the solubility of alloy elements occurs to a lower extent in Nb-rich MC and the hardness is closer to that of the pure carbide, ~2400 HV [11].

In addition to the hardness, the size and distribution are critical for abrasive wear. The size, schematically shown in the second column of Figure 5.5b, has to always be evaluated in comparison to the size of the abrasive agent. But, in general terms, it may be affirmed that carbide sizes smaller than 2 μm may not bring in a strong effect, being easily removed by the wear groove (Figure 3.27b). An example of this is displayed in Figure 5.6a for a powder metallurgy (PM) high-V, high-C tool steel. However, with a high volume of very hard carbides (of MC type), those are easily removed away and the net effect in wear resistance is small. In fact, when the abrasive size decreases, the wear resistance of this steel increases, surpassing that of D2, which is a sign that the small size of the PM tool steel carbide leads to a better wear resistance when the abrasive agent is also small. On the other hand, when carbides are badly distributed, such as in Figure 5.6b, the abrasive wear may also take place at areas that are not covered by carbides, as if those areas were not "protected" against abrasion. The ideal situation for abrasive resistance would be large and well-distributed carbides, as shown in Figure 5.6c as an example by production via the spray-forming technique SF, although this is not a common casting method for commercial products. Therefore, the best abrasive resistance is usually found in steels with a large carbide content but with good particle distribution (Figure 5.6d); considering existing commercial products, this microstructure is produced by conventional casting but with a combination of adequate ingot size and deformation. For more discussion, see Section 5.3.2.

5.2.2.2 Adhesive Wear

It is hard to find in the literature a clear definition of adhesive wear, but it is commonly seen in a metal on metal wear combinations, although it may be observed in other types of materials as well. Adhesion occurs when two similar materials are placed in close contact in a way that the atoms are able to develop chemical bonds. For this reason, it has been demonstrated that the more similar are the materials, the more likely adhesion occurs, with the highest adhesion and friction forces observed for friction between the same material [17]. In a more practical view, the literature often describes adhesion by the term "cold welding." In order to enable this bonding, the natural oxide layer of two metals must be broken, which usually occurs when high pressure is applied and both metals are plastically deformed, thereby coming into very close contact. If the temperature is high, surface oxidation may prevent adhesion once an oxide layer is continuously formed between the surfaces. Therefore, considering all those conditions, cold work tooling may be an ideal condition for adhesion, such as stamping and fine blanking or other operations: the tool steel usually forming another steel sheet or blank (of the same material) under high pressure (plastic deformation) and at room temperature (no surface oxidation). Cases of adhesion may also be seen in high-speed, high-temperature forming, where oxidation of the surface is prevented by the close contact between the tool and the workpiece, a condition typically observed in metal machining with high-speed steel or hard metal tools.

In most cases, adhesion occurs at small points, as shown in Figure 5.7a, and wear occurs when the interfaces in contact are made to slide and the locally adhered regions must separate [18]. In tooling operations, when parts are formed and extracted from the dies, the adhered areas may cause removal of pieces from the tool, or adhesion of the small areas of the workpiece occurs on the tool surface, both of which cause dimensional variations on the tool and impair the tooling

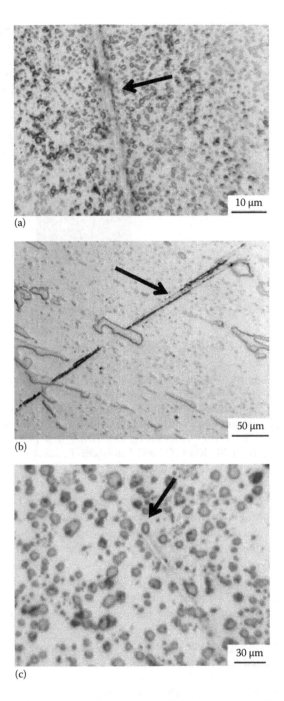

(a)

(b)

(c)

FIGURE 5.6 Example of surface damaged in polished specimens, after abrasion. (a) PM 10%V 2.6%C steel. (b) D6 tool steel (similar to D3). (c) Spray-formed 10%V 2.6%C steel. (From Mesquita, R.A. and Barbosa, C.A., *Proc. Euro PM 2004 (Pow. Metal. World Congr. Exhibit.),* 5, 53.)   *(Continued)*

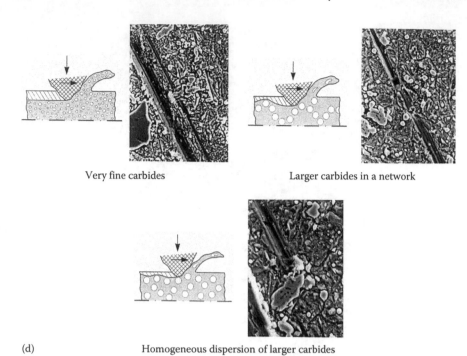

Very fine carbides   Larger carbides in a network

(d)   Homogeneous dispersion of larger carbides

FIGURE 5.6 (*Continued*)   Example of surface damaged in polished specimens, after abrasion. (d) Schematic comparison showing the effect of different carbide distributions, the situations being presented similar to the micrographs shown in (a–c). (From Schruff, I. et al., Advanced tool steels produced via spray forming, *Proceedings of the Sixth International Tooling Conference—The Use of Tool Steels: Experience and Research*, Bergstrom, J., Fredriksson, G., Johansson, M., Kotik, O., and Thuvander, F., eds., pp. 1159–1179, Karlstad, Sweden, 2002.)

performance. Often in the literature, the first possibility, that is, when the tool is damaged by the adhesion and loses its dimension, is named *adhesive wear*. On the other hand, the second possibility—the combination of adhesion and building up of soft steel on the surface—is referred to as *galling*.* Examples of adhesive wear and galling are shown in Figure 5.7b and c, respectively. For simplicity, throughout this book, all kinds of tool failures in tooling related to adhesion will be referred to as adhesive wear (Figure 5.7), since the mechanism is always the same, with metallic bonds between the tool and the formed piece, just varying where the position of the built-up damaged material.

The possibilities of avoiding adhesive wear may be achieved by understanding the mechanisms described earlier, by five main means:

1. Avoid the points of contact with high concentration of stresses, such as by changes in the design, better tool polishing, and the removal of grinding marks.

---

* The broken micropieces of both tool steel of the work-piece may oxidize and lead to abrasive wear, causing a mixture of adhesive and abrasive wear failure.

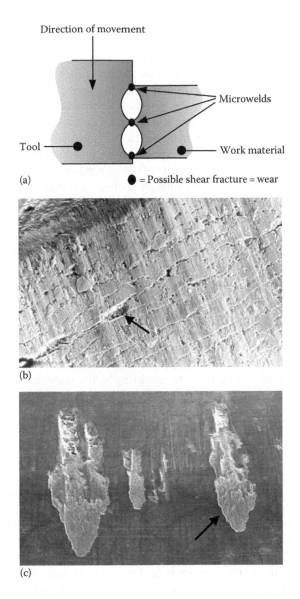

(a)

(b)

(c)

FIGURE 5.7 Schematic representation of adhesion and damage caused by it. (a) Points where the high pressure between the two metal surfaces may cause adhesion, which will lead to subsequent fracture of one of the materials during the tool movement. (From Uddeholm Tool Steel for Cold Work Tooling, Available at: http://www.uddeholm.com/files/, Accessed January 2016.) (b) Material loss at the tool surface, when adhesion is combined with friction (relative movement between the two surfaces). (From Uddeholm Tooling, Technical catalogue on vancron tool steel, Available at: www.uddeholm.com/files/PB_vancron_40_english.pdf, Accessed in March 2015.) (c) Example of building up the soft material on the tool surface, also called *galling.* (Also from Uddeholm Tooling, Technical catalogue on vancron tool steel, Available at: www.uddeholm.com/files/PB_vancron_40_english.pdf, Accessed in March 2015.) Images at (b) and (c) are with unknown magnification, but it is probably between 50× and 100×.

2. Create an intermediate layer between the work-piece and the tool steel by adding or improving the lubrication.
3. Place a ceramic layer on top of the tool steel, which may be done by the application of PVD coatings based on carbides or nitrides, and so with distinct bonding characteristics than ferrous alloys.
4. Increase the amount of carbides and their distribution, which will also act as areas with distinct chemical characteristics and so less prone to adhesion. Improvement of carbide distribution also helps in increasing the toughness by avoiding microcracking damage that often follows the adhesion.
5. Increase the tool hardness, which will reduce plastic deformation and then reduce adhesion. Surface treatments such as nitriding may help in this regard, in addition to the possibility of adding a nonmetallic compound layer and then with less adhesion.

Therefore, in terms of tool steel microstructure, the best performance in adhesive tooling conditions are found when a large number of fine carbide particles are dispersed in a high-hardness matrix, typically between 60 and 65 HRC. Those conditions are especially observed in powder metallurgy materials, with high carbon (>1%C) and high amounts of strong carbide-forming alloying elements (>5% alloy content), especially V, Mo, and W but also Cr. For conventional materials, the D-series are often used in adhesive conditions, and grades with proper distribution of carbides usually show better results (see Section 5.3). The importance of the fine distribution of carbides, in addition to providing higher toughness, is pointed as the main reason for the better performance of 8%Cr in sheet-forming adhesive conditions when compared to traditional D-series steels [20].

5.2.3 TOUGHNESS AND CRACK PROPAGATION

Bringing back the discussion from Figure 5.3, in addition to the increase in yield strength, hardness leads to a strong decrease in ductility, that is, the ability to deform plastically before fracture. So, once the yield point (initial of plastic deformation) is crossed, tool steel will fracture rapidly. The area under the stress–strain curve—which mathematically represents the strain multiplied by stress—gives the energy absorbed through fracture and thus is a good indication of the material toughness. Following this rationale, the increase in hardness is then expected to generate a dramatic decrease in toughness, which indeed occurs in cold work tool steels (and high-speed steels). This drawback follows from the same traditional compromise between strength and toughness that exists in many metals but is more critical in cold work tooling due to the extremely high hardness (extremely high strength) and so the very low corresponding toughness.

Figure 5.8a shows a comparison of the impact toughness of cold work tool steels, which is much lower than values of up to 20 J (Charpy V) observed in hot work tool steels. In fact, several failures of premature cracking may exist in cold work tool steels, such as those shown in Figure 5.9, especially in areas where the stresses are concentrated (also called *stress risers*), such as sharp corners and manufacturing marks. By not being able to deform plastically, due to the high hardness, cracks will

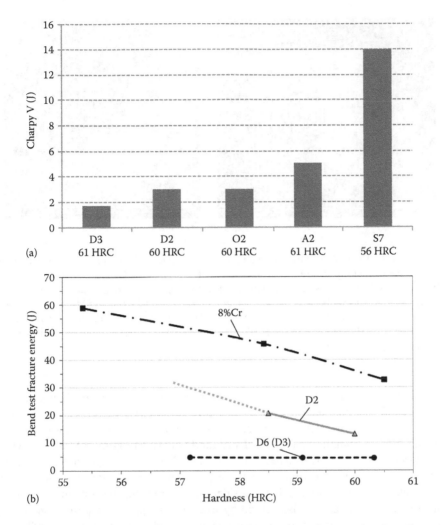

FIGURE 5.8 (a) Impact toughness of several cold work steels. Observe that the values are quite low and for this reason the Charpy impact test (V-notched) is often substituted by others, such as unnotched impact test. (Adapted from Hemphill, R.M. and Wert, D.E., Impact and fracture toughness testing of common grades of tool steels, in: *Tool Materials for Molds and Dies*, Krauss, G. and Nordberg, H., eds., Colorado School of Mines Press, Golden, MO, 1987, pp. 66–69.) (b) Comparison of toughness of D-series and 8%Cr cold work steels, with different hardness levels, measured by the bend test (specimens of 5 × 7 mm²). (From Mesquita, R.A. and Barbosa, C.A., *Technol. Metal. Mater.*, 2, 12, 2005; also used in Figure 3.28.)

emerge and propagate fast through the material. So, special care should be taken in the design of cold work tools, avoiding those stress risers, and also avoiding excessive stresses, either during quenching in heat treatment or during tool usage.

Again, in the case of the new generation of 8%Cr steels, toughness is considerably higher for those grades (Figure 5.8b) even when comparing different hardness levels.

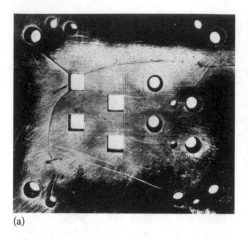

(a)   (b)

FIGURE 5.9 Example of failures related to the low toughness of cold work steels, which leads to easy cracking in stress concentration spots. (a, b: From Canale, L.C.F. et al. (eds.), Failure analysis in tool steels, in: *Failure Analysis of Heat Treated Steel Components*, Mesquita, R.A. and Barbosa, C.A., eds., American Society for Metals (ASM), vol. 1, Chapter 11, 2008, pp. 311–350; After Sullivan, J.W., *ASM Met. Eng. Q*, 13, 31, February 1973; Davis, J.R. (ed.), *Tool Materials*, ASM Specialty Handbook, ASM International, Materials Park, OH, 1995, pp. 421–441.)

The first reason is related to the high-temperature tempering, explained earlier (Section 5.2.1), which leads to a microstructure with better stress relief on the martensite structure. However, a second reason also exists, in terms of the lower amount of carbides, also smaller and better distributed (shown in Figure 5.14 ahead) when compared to D-series steels (check also Figure 3.25). While important for abrasive wear resistance, those carbides constitute preferential points for crack nucleation and routes for crack propagation, considerably decreasing the toughness. In fact, since adhesive damage is also caused by microcracking, several results show that 8%Cr steels tend to present better wear resistance and performance under adhesive conditions [20].

In conclusion, while cold work tool steels are designed to offer high hardness (resistance against plastic deformation) and wear resistance, toughness is naturally low. Options such as the 8%Cr do exist, but cold work steels will still be in a very brittle condition, and (with exception of the S-series designed for shock-resistant applications) the consequences for this low toughness should be taken into account in the design and usage of a cold-forming tool.

5.3 MAIN MICROSTRUCTURAL FEATURES IN COLD WORK STEELS: CARBIDES AND RETAINED AUSTENITE

Following the approach of this book of understanding the properties and performance based on the microstructure, the main microstructural features of cold work tool steels are considered in this section. The previous sections showed that high strength—typically referred to in practical applications as high hardness—and high wear resistance are the two main properties of cold work steels. Here, the ways to achieve those properties are considered in detail by the microstructure of high-carbon tempered martensite and the dispersion of alloy carbides.

5.3.1 MARTENSITE AND RETAINED AUSTENITE IN COLD WORK STEELS

5.3.1.1 Carbon Content and Martensite Hardness

Although many possible strengthening methods exist for steels, the combination of high carbon and the hardening process, by austenitizing and quenching, is by far the most powerful in achieving high strength, leading to hardness increase beyond 60 HRC or more than 3000 MPa. The main reason is the effect of carbon on martensite hardness, which is basically constituted by fine acicular grains with a very high density of dislocations. When the carbon content increases, the combination of supersaturated carbon in solid solution and the high density of dislocations leads to a tremendous increase in hardness, as shown in Figure 5.10. However, after 0.40%C, the rate of hardness increase by carbon additions to martensite decreases, achieving a maximum at 0.8%C.

For higher carbon contents, there is actually a decrease in the hardness (Figure 5.10), which cannot be explained by the carbon effect in martensite but by the presence of the austenite phase that is not transformed to martensite, named *retained austenite* (also described in Section 3.2.2.2). Different from martensite, austenite is a soft phase and the increase in its amount decreases the hardness of the quenched steel.

One disadvantage of high-carbon steels and their ability to increase hardness after martensite transformation occurs when those steels, or at least small areas of them, are accidentally heated above the austenite temperature. Upon cooling, they may transform to martensite, causing brittleness and or even cracks during transformation on cooling. In practical terms, this may occur when tool steels are manufactured by grinding or by electrodischarge machining (EDM), as shown in Figure 5.11a and b, respectively. Both processes can cause surface damage to any kind of tool steels, but the problems tend to be more serious in cold work steels because of the high-carbon effect mentioned previously and because of the natural brittleness of those grades (Figure 5.8a).

Starting with grinding (example in Figure 5.11a), it is often applied for small changes and low amount of material removal, typical for finishing or sharpening of tools. Going more in depth about grinding, it is typically an abrasion mechanism, and since cold work tool steels are resistant to abrasion, their grindability is often limited, and, if excessive stress is applied during grinding, aiming at faster material removal, the result may be surface heating. If temperatures above A1 are achieved (entering the austenite field), the surface layer will transform after cooling to high-hardness martensite, which often contains microcracks. And when in use, these microcracks may grow and cause large cracking or fracture of the tool. In addition, the areas adjacent to the rehardened layer will have been heated above the normal tempering temperature, leading to a hardness loss and also heterogeneities

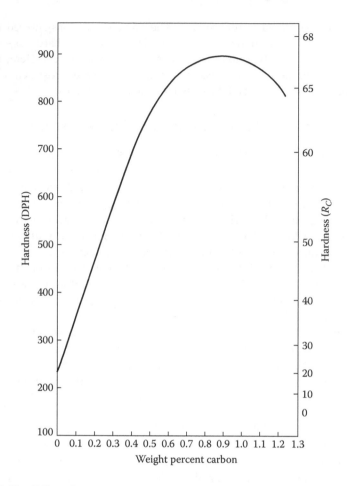

FIGURE 5.10 Effect of carbon on the hardness of unalloyed quenched steels. (From Krauss, G., Martensitic transformation, structure and properties in hardenable steels, *Hardenability Concepts with Application to Steel*, TMS-AIME, 1978, pp. 229–248.) DPH means diamond pyramid hardness.

in the microstructure. The final result in the areas affected by overheating during grinding will be similar to the hardness profile shown in Figure 5.11a: a surface layer harder than expected, due to hardening after grinding, followed by an area with lower hardness than expected, due to overtempering. To avoid this, it is always recommended to use low stock removal parameters during grinding, especially when dealing with highly abrasive resistant cold work tool steels such as D2 and mainly D3 and D6. And, after grinding, a new tempering, at ~40°C (80°F) below the tempering temperature used during the tool heat treatment, is applied, which will temper any remaining fresh martensite.

The damage to the tool surface by EDM, in metallurgical terms, is similar to that of grinding, previously discussed. During EDM, an electric discharge heats small areas on the tool surface, which are then removed and dissolved in the electrolyte.

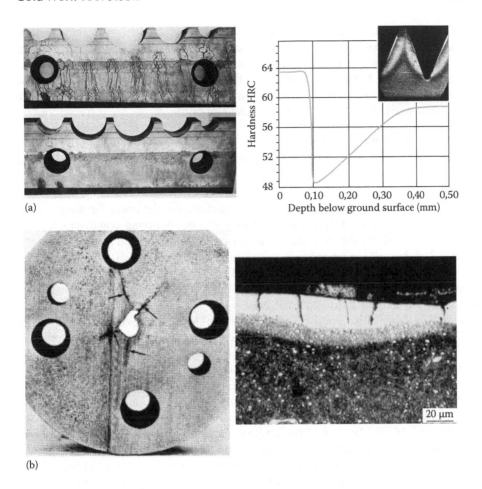

(a)

(b)

FIGURE 5.11 Surface overheating during (a) grinding and (b) EDM (electrodischarge machining). In both cases, the surface regions are heated and transformed to martensite, causing cracks in the fresh martensite areas (lighter than the normal matrix, in the micrographs) and lower hardness in the adjacent areas (darker than the normal matrix). (a: From Davis, J.R. (ed.), *Tool Materials*, ASM Specialty Handbook, ASM International, Materials Park, OH, 1995; Uddeholm Tooling, Grinding of tool steel, Technical brochure, Available at: www.uddeholm.com/files/TB_grinding-english.pdf, Accessed March, 2016; b: From Canale, L.C.F. et al. (eds.), Failure analysis in tool steels, in: *Failure Analysis of Heat Treated Steel Components*, Mesquita, R.A. and Barbosa, C.A., eds., American Society for Metals (ASM), 2008, vol. 1, Chapter 11, pp. 311–350; ASM Handbook, *Failure Analysis and Prevention*, 8th edn., vol. 10, American Society for Metals, Materials Park, OH, 1975, pp. 500–507.)

However, if high removal rates are used, areas of the heated surface will remain, which may form a rehardened layer on top of the tools, very often resulting in small cracks (Figure 5.11b). In addition to the martensite transformation, the reason for those cracks is also the existence of remelted and resolidified areas, with a very coarse microstructure and often with porosities. One solution would be to avoid EDM and use a combination of high-speed machining and grinding to produce the tools. But, in many cases EDM is still necessary, being then recommended that the parameters suggested by the EDM manufacturer are not exceeded, and they are even reduced further during the last steps of material removal. This will guarantee a small affected layer by EDM at the end. Lastly, the tools should be retempered, so that the remaining as-hardened microstructure on the rest of surface layer is converted to tempered martensite, which is less hard and less brittle.

5.3.1.2 Retained Austenite

The effects in Figure 5.10, explaining why carbon increases the amount of retained austenite, may be understood by analyzing Figure 5.12a through d for a 1%C 2.8%Cr steel. In Figure 5.12a, it is seen that, under cooling, martensite amount starts to increase as soon as M_s is reached, and the lower the temperature, the higher the martensite content. Under some austenitizing conditions (e.g., at 840°C or 1540°F), the martensite transformation is completed at room temperature, and thus all the austenite is transformed into martensite (when temperature reached Mf).* However, for higher austenitizing temperatures (e.g., 930°C/1700°F and 1040°C/1900°F), this does not occur, and a certain amount of untransformed austenite (retained austenite) will exist when the steel reaches room temperature. The same effect is observed by increasing the carbon content, as shown in Figure 5.12b: the higher the carbon content, the larger the amount of retained austenite. Both effects, of the carbon content and hardening temperatures, may be understood by taking into consideration the chemical composition effect on the martensite transformation temperature. Again, in Figure 5.12a, it may be observed that M_s and M_f temperatures tend to "move together," meaning that if the M_s temperature is lowered, the M_f temperature will also be lower. As a result, when M_s decreases, M_f will be shifted further down, and a higher portion of austenite will not be transformed and so the amount of retained austenite will increase. In this respect, the effect of all elements, including carbon, on M_s is shown in Figure 5.12c. As a rule, carbon has a strong effect in reducing M_s and M_f temperatures, and so have all alloying elements, with a few exceptions such as cobalt and aluminum.

In summary, the discussion on the results in Figure 5.12 points that, whenever carbon and more alloying elements are placed in solid solution, the M_s temperature will decrease and the amount of retained austenite will increase. This occurs basically in two ways: increasing the amount of carbon and alloy elements to the overall steel composition, or increasing the amount of those elements in solid solution, the last factor being deeply related to the hardening temperature used. The higher the hardening temperature, the higher the dissolution of carbides, and then the higher the amount of elements in the solid solution, leading to an increase in the retained austenite. This is very important to cold work tool steels (and also to high-speed steels); some results for three different steels are shown in Figure 5.12d. For D2 steel, for example, higher austenitizing temperatures lead to more than 80% of the

---

* The temperatures M_s and M_f are defined as the start and final temperatures for martensite transformation, as explained in Section 3.2.2.2 of Chapter 3. Martensite start to be formed during quenching as soon as non-transformed austenite reaches M_s. The amount of martensite increases with cooling down, leading to a 100% martensitic structure when Mf temperature is reached. If M_f is below room temperature, retained austenite will then be present after quenching.

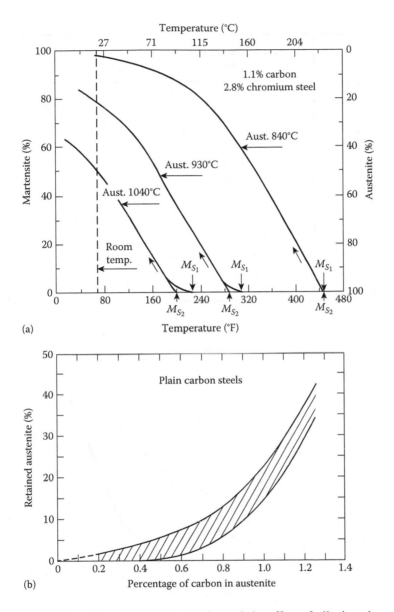

(a)

(b)

FIGURE 5.12 Formation of retained austenite and the effect of alloying elements. (a) Transformation of austenite to martensite upon cooling from different hardening temperatures. Depending on the conditions, austenite at room temperature will not be fully transformed to martensite, thus generating the microconstituent known as retained austenite. (b) Effect of carbon on the increase of retained austenite. (Charts from (a) to (c) as well as the A4 steel data used in item (d) were all obtained from Cohen, M., *Trans. ASM*, 41, 35, 1949; In (d), A2 steel data is from Averbach, B.L. et al., *The Effect of Plastic Deformation on Solid Reactions, Part II: The Effect of Applied Stress on the Martensite Reaction*, Cold Working of Metals, American Society for Metals, 1949; and D2 data from Zmeskal, O. and Cohen, M., *Trans. ASM*, 31, 380, 1943.)   (*Continued*)

Element	Relationship between the effects of elements on M_s and on propensity for retaining austenite	
	Change of M_s per 1% of element	Change of retained austenite per 1% of element (in the presence of 1%C)
Carbon	−300°C (−540°F)	+50%
Manganese	−33°C (−60°F)	+20%
Chromium	−22°C (−40°F)	+11%
Nickel	−17°C (−30°F)	+10%
Molybdehum	−11°C (−20°F)	+9%
Tungsten	−11°C (−20°F)	+8%
Silicon	−11°C (−20°F)	+6%
Cobalt	+6°C (+10°F)	−3%
Aluminum	+17°C (+30°F)	−4%

(c)

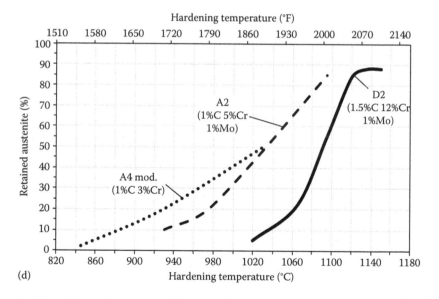

(d)

FIGURE 5.12 (Continued) Formation of retained austenite and the effect of alloying elements. (c) Approximate quantitative relations for carbon and other alloying elements on the M_s temperature and on the amount of retained austenite. (d) Effect of hardening (austenitizing) temperature on the amount of retained austenite of different cold work steels. (Charts from (a) to (c) as well as the A4 steel data used in item (d) were all obtained from Cohen, M., *Trans. ASM*, 41, 35, 1949; In (d), A2 steel data is from Averbach, B.L. et al., The effect of plastic deformation on solid reactions, Part II: The effect of applied stress on the martensite reaction, *Cold Working of Metals*, American Society for Metals, 1949; and D2 data from Zmeskal, O. and Cohen, M., *Trans. ASM*, 31, 380, 1943.)

microstructure composed of retained austenite, which is almost the whole micro-structural matrix, since ~10% is composed of undissolved carbides.

Another undesirable effect of retained austenite, in addition to the decrease in hardness, is the possibility for retained austenite to transform to martensite under application of stresses, even if the M_f temperature is not reached. This means that the stresses developed during cold form tooling may lead to martensite formation, which in turn may cause the expansion of the microstructure and then lead to dimensional instabilities of tool steels. In fact, in many fine-blanking or other high-precision cold work tools, it is common to use steels with a low amount of retained austenite to eliminate dimensional instabilities. In addition, the austenite transformed by mechanical stress may be as brittle as fresh martensite, and so may lead to small cracks or chipping in the forming areas; this then points to a second important reason to avoid expressive amounts of retained austenite.

5.3.1.3 Reduction or Elimination of Retained Austenite

One way to reduce or eliminate the amount of retained austenite is by tempering at high temperature, typically above 500°C (930°F). As shown in Figure 5.13, on an example for D2 tool steel, there is a continuous decrease in retained austenite by increasing the tempering temperature above 400°C (750°F), with retained austenite being practically eliminated after tempering above 500°C (930°F). The reason, as explained in Section 3.2.2.2, is the precipitation of carbides during tempering, which decreases the amount of carbon and alloy elements in solid solution in austenite and then raises M_s and M_f temperatures, promoting the transformation of retained austenite to martensite upon cooling to room temperature from the first tempering. And then a second tempering is necessary to temper the fresh martensite resulting from the transformation of retained austenite after the first tempering. So, for high-alloy cold work steels (e.g., from A, D, or even O series), two tempering treatments are always recommended.

High tempering temperature is also important when nitriding or other surface hardening or coatings are applied, and then the tools are exposed to temperatures above 400°C/500°C (750°F/930°F). So, to improve the secondary hardening during tempering, hardening temperatures may be raised, especially for D2 steel. While this is a solution to stabilize the hardness, the increase of retained austenite and the grain growth during austenitizing may limit the increase in hardening temperatures. For example, there are many papers considering the possibility of increasing the hardening temperatures in D2 to achieve 60 HRC after tempering at ~550°C (1020°F) (compare Figure 5.13a and b). While this is successful in increasing the stability during nitriding or other surface hardening methods, it is shown in several cases that the decrease in toughness and microstructure instabilities is due to the excessive amount of retained austenite.* It is the opinion of this author that the best practice would be

---

* Although the tempering charts show that the elimination of retained austenite is possible by high-temperature tempering after of high hardening conditions (Figure 5.13c), practical situations may not be as simple. As detailed in [28], retained austenite may be established after a few hours at room temperature, leading to no transformation. This is quite common in large industrial tools, which take quite a long time to reach room temperature and when the practices do not allow immediate tempering after hardening. So, if the amount of retained austenite is very high, for example, exceeding 50%, it may not be fully eliminated and then the deleterious effects of the retained austenite will remain in the final tools.

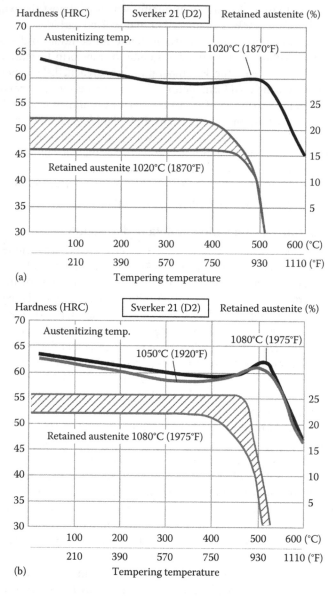

FIGURE 5.13 Tempering curve for (a), (b) D2 and (c) 8%Cr tool steel, showing the effect of increasing the tempering temperature on the decrease of hardness and also on the reversion of retained austenite. (Adapted from Uddeholm Tooling, Product Brochure Uddeholm Sverker 21, Available online at: http://www.uddeholm.com/files/PB_Uddeholm_sverker_21_english.pdf, Accessed February 2016; Uddeholm Tooling, Product Brochure Uddeholm Sleipner, Available online at: http://www.uddeholm.com/files/PB_Uddeholm_sleipner_english.pdf, Accessed February 2016.) Observe that for the higher hardening temperatures (b), D2 achieves higher hardness after high-temperature tempering, but also higher levels of retained austenite. Observe also that lower C leads to smaller amounts of retained austenite, which is completely eliminated after tempering at 550°C (1020°F), still maintaining 60 HRC.   (*Continued*)

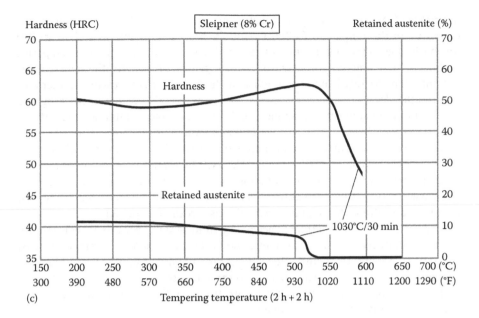

FIGURE 5.13 (*Continued*) Tempering curve for (a), (b) D2 and (c) 8%Cr tool steel, showing the effect of increasing the tempering temperature on the decrease of hardness and also on the reversion of retained austenite. (Adapted from Uddeholm Tooling, Product Brochure Uddeholm Sverker 21, Available online at: http://www.uddeholm.com/files/PB_Uddeholm_sverker_21_english.pdf, Accessed February 2016; Uddeholm Tooling, Product Brochure Uddeholm Sleipner, Available online at: http://www.uddeholm.com/files/PB_Uddeholm_sleipner_english.pdf, Accessed February 2016.) Observe that for the higher hardening temperatures (b), D2 achieves higher hardness after high-temperature tempering, but also higher levels of retained austenite. Observe also that lower C leads to smaller amounts of retained austenite, which is completely eliminated after tempering at 550°C (1020°F), still maintaining 60 HRC.

to use increased tempering but not exceeding the hardening temperature; this leads to hardness between 55 and 58 HRC and elimination of retained austenite. However, there are other possibilities when the 8%Cr steels are used in substitution to D-series steels. In addition to higher toughness, 8%Cr steels have lower C content as well as high precipitation hardening characteristics, mainly due to higher Mo amounts, which combined leads to lower amounts of retained austenite and its elimination during high-temperature tempering (compare Figure 5.13b and c), still maintaining the hardness levels of ~60 HRC.

Another way to reduce or eliminate the amount of retained austenite is by cooling the tools, after quenching, down to temperatures close to or below the martensite finish temperature (M_f). These treatments are known as *cryogenic treatments* and may be primarily applied to tools that are sensitive to dimensional changes and that cannot be tempered at high temperature, involving hardness decrease, for example. The application of cryogenic treatment, directly after tempering, is, however, tricky in industrial samples, because the stresses of martensite transformation may lead to cracking.* So, some cryogenic treatments are preceded by a low-temperature tempering in the range 100°C–200°C (210°F–390°F). This, however, has another drawback, due to the stabilization of retained austenite during this low tempering treatment. So, in practical terms, the use of cryogenic treatment to eliminate retained austenite may be restricted. There exist, in fact, several sources in the literature pointing that long cryogenic treatments far below M_f (also called deep cryogenic) may lead to improvement of toughness and wear resistance, claiming at significant microstructural changes (such as carbide precipitation) as the main cause. However, scientific explanations and the confirmation of measured properties are still not clear, and it is possible that the effects are only caused by the transformation of retained austenite and, in the case of low tempering prior to cryogenic treatment, also related to the stabilization of retained austenite (see discussion in [33]), which may increase the toughness (although bringing drawbacks, as already mentioned).

As a conclusion, high carbon is necessary in most cold work tool steels, but the detrimental effects of retained austenite should be controlled, in terms of decrease in hardness, possibility of embrittlement, and mainly dimensional instabilities. The mechanisms to do so are related mainly to three points: the composition of the tool steel used, the hardening temperature (controlling the dissolution of alloy carbides), and the tempering temperature applied.

5.3.2 Overview on Carbides in Cold Work Tool Steels and the Effects of Carbide Distribution

5.3.2.1 Chemistry and Hardness of Carbides

As explained in Section 5.2.2, undissolved carbides are essential to the wear resistance of tool steels. In abrasive wear, large carbides are usually desirable, and for adhesive wear small carbides are preferred, while for both mechanisms a good distribution of carbides is important. The first possibility for undissolved carbides would be the addition of carbon and the formation of iron carbides (cementite), such as in white-cast irons. However, the hardness of cementite is not enough to promote adequate wear resistance in tooling conditions, and so carbides of alloying elements are used. In this sense, Table 5.2 show the chemical, structural, and hardness comparison of the most common undissolved carbides applied in cold work tool steels.

The first option of alloy carbides in cold work tool steels, concerning cost and hardness, would be the addition of Cr-rich carbides of the M_7C_3 type. Other alloy carbides may be used, promoting higher hardness but involving alloy elements that are more expensive: W, Mo, V, or Nb. As explained in Figure 3.12, those elements have to be added in amounts higher than their solubility, which then will lead to carbides that are not dissolved during hardening and are mainly formed directly after solidification (Figure 3.28). This means roughly (for 1%C) more than 3%W, 3%Mo, and 0.5%V, while Nb will form carbides in any amounts above 0.01%Nb. So, the addition of those alloying elements would involve cost increase, especially

---

* When austenite is present, this phase can absorb the stresses due to the high ductility. But, after full transformation to martensite, this is no longer possible and cracks may easily occur.

TABLE 5.2

Data for Undissolved Carbides Typically Found in Cold Work Tool Steels

Phase	Rich in	Pure Carbide (Reference)	Crystallographic System	Lattice Parameters (Reference)	Hardness (Reference)
MC	Nb	NbC	Face centered cubic	$a = 4.47$ Å [37]	2300 HV [44]
MC	V	V_4C_3	Cubic	$a = 4.16$ Å [38]	2000 HV [45]
M_2C	Mo	Mo_2C	Hexagonal	$a = 3.01$ Å; $c = 4.74$ Å [39]	1800 HV [46]
M_6C	W or Mo	Fe_3Mo_3C	Cubic	$a = 11.12$ Å [40]	1500 HV [47]
M_7C3	Cr	Cr_7C_3	Hexagonal	$a = 13.9$ Å; $c = 4.52$ Å [41]	1600 HV [48]
M_3C	Fe	Fe_3C	Orthorhombic	$a = 5.06$ Å; $b = 6.74$ Å; $c = 4.50$ Å [42]	1100 HV [49]
Martensite	0.8%C steel	—	Tetragonal	$a = 2.85$ Å; $c = 2.95$ Å [43]	900 HV (65 HRC)

Note: References shown by each of the data.

considering the addition above the solubility limits of each element. Mo and W carbides are then commonly seen only in high-speed steels, while some tool steels would present high V to improve wear resistance. In the vast majority, however, cold work tool steels are based on Cr carbides, by presenting high amounts of Cr (between 5% and 12%) and C (1%–2%) contents. More details on those carbides are also discussed in [34].

In this sense, NbC is shown as an interesting possibility, because of the lower solubility and then the straight formation of carbides after small additions. Another advantage of NbC is the very high hardness of this carbide, which is the hardest in Table 5.2. In spite of that, Nb is not used extensively in tool steels, mainly because the development of most tool steels used today was done before the 1960s when Nb was not an element as available as today. So, it is the opinion of this author that the use of Nb as a carbide-forming element would represent an important potential. In fact, several new developments show that this element can be successfully applied in high-speed steels [35,36], and the possibilities in cold work would be a direct approach. In fact, the cold work tool steels used today involve several concepts that were developed from high-speed steels.

5.3.2.2 Amount and Distribution of Carbides in Cold Work Tool Steels

In order to achieve the expected wear resistance, not only the type but also the size and the distribution of carbides are important, as explained in Section 5.2.2. In this section, examples and a deeper discussion on both aspects are given. Starting from the amount of carbides, Figure 5.14 gathers examples of the most important steels used in cold work and their respective microstructure. The gradual decrease of the amount and size of large carbides from D3 to D2, from D2 to 8%Cr and from 8%Cr to A2, is clear. The reason is basically due to the amount of C and Cr in those

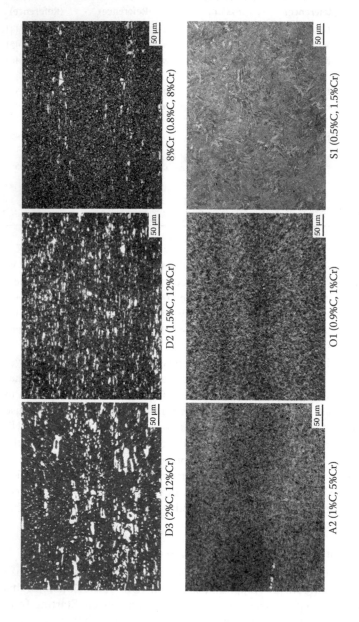

FIGURE 5.14 Microstructure of typical cold work steels, from mid-radius of bars of 60–90 mm diameter, all in hardened and tempered conditions for the most typical hardness (58–60 HRC for all steels except for S1, which is heat-treated to around 55 HRC). (D3 and O1 from Mesquita, R.A. and Barbosa, C.A., *Technol. Metal. Mater.*, 2, 12, 2005; D2 and 8%Cr from Uddeholm Tooling, Product Brochure Uddeholm Sleipner, Available online at: http://www.uddeholm.com/files/PB_Uddeholm_sleipner_english.pdf, Accessed February 2016; A2 and S1 from the author.) Remarks: (1) The amount of Cr and C represent the typical for those grades; (2) For reference, also check Figure 3.25; (3) For D2, the microstructure considers tempering in higher temperature (~550°C or 1020°F), while in Figure 3.25, D2 was tempered at a lower temperature (~250°C or 480°F).

grades once those elements are combined to form M_7C_3-type carbides (particularly the product %C·%Cr. When both C and Cr contents increase, the amount of carbides that cannot undergo dissolution in solution also increases, leading to a higher amount of undissolved carbides. For similar reasons, O1 and S1 practically do not present undissolved carbides in their microstructure, the combination of Cr and C being too low to enable complete solid solution of those elements after solidification. The difference observed in the microstructure of O1 and S1 is not caused by the carbides but related to the matrix composition, especially the carbon content (~1% in O1 and 0.6% in S1 tool steel).

A gradual decrease in the wear resistance, especially under abrasive conditions, is expected in the same order of the decrease in amount of undissolved carbides in Figure 5.14. In the case of O1 and S1, the wear resistance will mainly depend on the hardness, which is higher for O1 due to its higher carbon content. For adhesive wear resistance (including galling), the distribution of carbides and toughness are also important, and usually D2 and 8%Cr steels show the best results.

More specifically, on the distribution of carbides, it was explained in Section 3.3.3 that the eutectic and primary carbides, which usually do not dissolve during hardening, have their distribution related to a combination of solidification and hot-deformation aspects. In this sense, not only the chemical composition, such as the %C and %Cr as shown in Figure 5.14, but also the production conditions will be important to define how the carbides will be placed in the microstructure, or, in other words, will determine the final carbide distribution. In this sense, charts like that shown in Figure 5.15 may be created for a specific type of tool steels, where the microstructures may be used as an approval criterion for the tool steel quality.* For this specific example, the microstructures are not approved when the carbide distribution is excessively coarse for a given dimension.

As also explained in Section 3.3.3, coarse carbide distribution occurs when either the as-solidified structure was excessively coarse or the amount of deformation applied in hot forming was not high enough. This in turn is given by the production parameters of the steel mills, which will define the ingot sizes and geometries as well as the processing conditions to achieve the final dimension (e.g., a combination of passes in hot-forging or hot-rolling). Therefore, this shows again the importance of not only the composition but also the production processes in the final quality of a given tool steel, as explained in Chapter 1 (Section 1.1) of this book.

In Figure 5.15, all microstructures were obtained from conventionally cast and forged/rolled products. However, when the casting process changes dramatically, meaning an increase of the cooling rate in solidification, the carbides may be dramatically changed. This is possible by casting methods alternative to ingot casting, when a high heat extraction is imposed to the liquid steel during solidification. The first method developed in this sense is powder metallurgy, or PM, as explained in Chapter 2 (Section 2.1.6). In this method, the liquid steel is not poured in a mold but is

---

* In a similar way, the North American Die-Casting Association and other quality standards use microstructures to determine the quality 5%Cr H-series hot work tool steels. In the case of H-series steels, however, the criterion is related to microsegregation and distribution of secondary carbides, as explained in Sections 3.2.1 and 4.2.2.

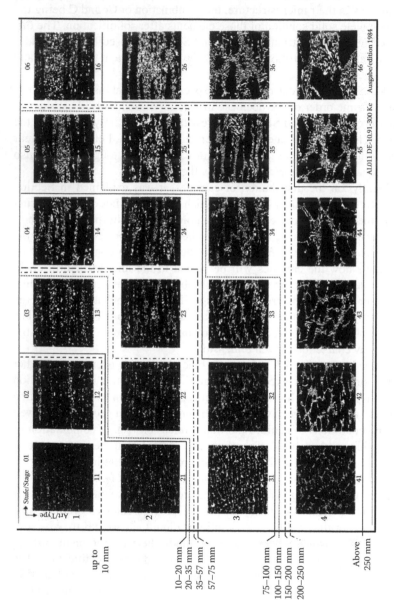

FIGURE 5.15 Distribution of carbides in D2 tool steel, depending on the dimensions. (According to Böhler Edelstahl, Reference charts for evaluating the carbide distribution in high carbon 12%Cr steels, printed in 1991, *Material Purchase Specification (Zukaufspezifikation Material)*, May, 2004.) The carbide distribution is better for the upper left areas and becomes worse as the microstructure moves to the bottom-right corner. The regions where the carbide distribution is approved are above the line for each dimension (samples taken from mid-radius and longitudinal direction). Microstructures here use deep etching on annealed structure, allowing more contrast between the carbides (white) and matrix (dark).

atomized: Droplets of liquid steel are created by a fine stream of liquid steel exposed to high-pressure gas, usually N_2. By doing this, the cooling rate during solidification increases by several orders of magnitude, generating much finer carbide particles. On the other hand, the final product is not a piece of steel, but a metal powder, with the same composition of the initial steel (often with high N contents, due to the N_2 atomization). To produce a bulk bar, the atomized powder is then pressed under high temperature and high pressure (usually the hot isotactic pressing, or HIP) into the form of a thick bar, which may be used right after pressing (condition named as-HIPed) when the HIP process is sufficient to eliminate the remaining porosity, or after the as-HIPed product is hot-rolled (References 51 and 52 explain this process in detail). The final result is carbides with much smaller sizes and very well distributed, which improve substantially the toughness, especially in the transverse direction, and adhesive wear resistance. Because the carbides become very small, the abrasive wear resistance of PM cold work and high-speed steels is usually smaller than that of the respective steels produced by conventional method. However, most cold-forming processes are related to adhesive wear and microcracking mechanisms, and so the PM grades usually promote substantial improvement in performance. An example of PM and conventional D2 is shown in Figure 5.16, illustrating the discussion above on carbide size, toughness, and wear resistance. However, it is important to note that PM is also a method to enable the production high-alloyed (high amount of carbides) tool steels once fine carbides can be easily hot-formed, and several tool steels show higher amount of alloying elements than D2, especially MC carbide forming element such as Nb or V.

Another casting process that is found in the literature is spray-forming (or SF). However, PM is much more established industrially than SF: for instance, most tool steel producers have several grades and regular commercial products with PM, while the existence of commercial products of SF is rather uncommon. Basically, SF has a similar atomizing method as PM, but the droplets are deposited on a surface before complete solidification. While the process is continuous, an ingot (also named *pre-form*) is grown, without the need for HIP. After finishing the deposition, ingots of several tons may be produced, with finer carbides (due to the partial solidification during atomizing) but still with porosity after consolidation. Those ingots are then forged to eliminate the porosity, and the final products are bars of tool steels, like the conventional cast products (see [54,55] for details on the SF process). In terms of toughness, the properties of SF are usually intermediary to those of conventional and PM, as shown in Figure 3.16 for D2. However, abrasive wear resistance of SF cold work tool steels to be superior to both PM and conventionally cast cold work steels, as shown in Figure 5.16c, since the carbides are large, an advantage in relation to PM, but also are better distributed when compared to conventional cast tool steels.

Considering only conventionally cast cold work steels, carbides will be forming coarse networks in the as-cast ingot, which are gradually broken during hot-forging and, for smaller dimensions (below 4 in., typically), hot rolling (see Figure 3.34). The breakdown of the carbide networks dramatically improves the properties, especially toughness and wear resistance, because of the improved carbide distribution. An example of this effect, also for D2 tool steel, is shown in Figure 5.17: The carbide networks are still clear in the large section of the bar, promoting large

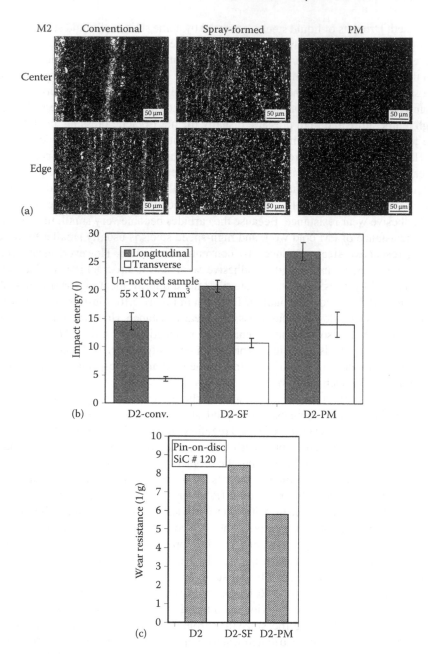

FIGURE 5.16 Variation of microstructure and mechanical properties of D2 tool steel cast with three different methods: conventional casting, spray-forming, and powder metallurgy. (a) Carbide distribution. (b) Toughness. (c) Abrasive wear resistance, pin-on-disc test on SiC. Size #120. (Adapted from Schneider, R. et al., The performance of spray-formed tool steels in comparison to conventional route material, *Proceedings of the Sixth International Tooling Conference—The Use of Tool Steels: Experience and Research*, Bergstrom, J., Fredriksson, G., Johansson, M., Kotik, O., and Thuvander, F., eds., pp. 1111–1124, Karlstad, Sweden, 2002.)

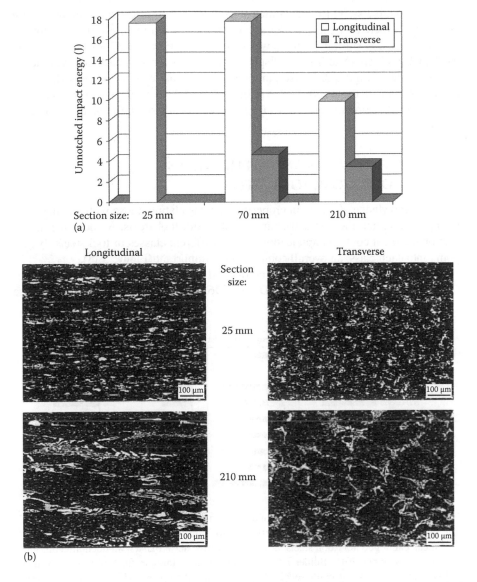

FIGURE 5.17 Effect of carbide distribution on the toughness of D2 tool steel. (a) Unnotched impact toughness ($7 \times 10 \times 55$ mm^3 specimens); longitudinal toughness for the 25 mm was not measured because the section size was not large enough to machine a specimen. (b) Microstructures for both the smallest and biggest section seizes. (From Mendanha, A. et al., Toughness of D2 cold work tool steel— Part I: The influence of the initial microstructure. *Proceeding of the Encontro de Integrantes da Cadeia Produtiva de Ferramentas, Moldes e Matrizes*, São Paulo, Brazil: ABM, October 2003. [In Portuguese, original title: tenacidade do aço ferramenta para trabalho a frio AISI D2—parte I influência da microestrutura de partida].)

agglomerates, which in turn create a preferential route for crack propagation and decrease toughness. Once the dimension decreases, that is, the amount of hot deformation increases, the carbides assume finer distributions, and then the toughness increases. A similar example was discussed in Figure 3.37 for an 8%Cr cold work tool steel. So, in addition to displaying the importance of carbide distribution, this example of Figure 3.17 also points to the fact that cold work tools should be always be made in dimensions close to the initial tool steel bar dimension, meaning that the construction of small tools from pieces taken from large blocks should be avoided.

5.4 SPECIFIC CLASSES AND APPLICATIONS OF COLD WORK TOOL STEELS

With the knowledge of the main properties and correlating those with the manufacturing method and the microstructure of cold work tool steels, we describe in this section the most common applications of the different classes of tool steels. Before going into each class, an overall comparison is important (Figure 5.18).

The discussion that follows will be divided into four groups. First is the high-Cr, high-C tool steels, comprising D2, D3/D6, 8%Cr, and A2 steels. In all these

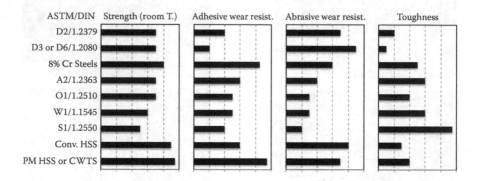

FIGURE 5.18 Qualitative comparison of the important properties of the main cold tool steels (the larger the bar, the higher the property). (Chart built with data from Uddeholm Tooling, Technical catalogue for tool steels, Available online at: http://www.uddeholm.com/files/, Accessed December 2015; Böhler Edelstahl, Technical catalogue for tool steels, Available online at: http://www.bohler-edelstahl.com/english/1853_ENG_HTML.php, Accessed December 2015; Villares Metals, Technical catalogue for tool steels, Available online at: http://www.villaresmetals.com.br/pt/Produtos/Acos-Ferramenta, Accessed December 2015.) *Notes:* (1) "Conv. HSS" means conventional high-speed steels applied in cold work tooling, such as M2 or M35 steel. "PM HSS" or "CWTS" means high alloy powder metallurgy produced high-speed steels and cold work tool steels, typically with high contents of V, Mo, and W. (2) Strength considers the achievable hardness after typical tempering conditions for each grade. (3) Wear resistance and toughness are also for the most typical hardness of each grade, which is ~58–60 for D, 8%Cr, A2, O1, and W1 steels, 55 HRC for S1, and 64 HRC for the high-speed steel and PM high-alloy cold work steels.

grades, the properties will be governed by the amount and distribution of Cr-rich M_7C_3 carbides, in addition to matrix effects related to the soluble alloying elements, mainly Mo and V, and the tempering conditions. The second group comprises the high-C, low-alloy steels O1 and W1, where a large amount carbide is not expected and the properties are governed by the high-carbon matrix. The third has very high alloy steels, such as conventional high-speed steels and high-alloy PM cold work tool steels. And, finally, we discuss steels that are much less wear- and shock-resistant, mainly the S1 grade.

5.4.1 High-Carbon High-Cr Steels: D-Series, A-Series, and 8%Cr Steels

D2 is the most relevant grade among the high-C, high-Cr cold work tool steels, and perhaps the most important reference for all cold work tool steels. It would not be an exaggeration to say that D2 is for cold work tool steels what H13 is for hot work tool steels. This is because this grade is very traditional, existing for several decades, and represents a good combination of properties such as wear resistance and toughness, in addition to a good balance in terms of cost. The 12%Cr content and 1.5%C lead to a large amount of Cr-rich carbides (Figure 5.14), which promotes high wear resistance. The amount is not as large as in the 2%C D-series steels, such as D3 or D6, promoting a better balance of toughness. The cost balance is given by the reasonably low amounts of expensive alloying elements, such as Mo and V, but enough to promote high hardenability (e.g., for gas quenching in vacuum processes) and secondary hardening after tempering.

Another grade close to D2 is D3 or D6 steels, the difference between the last two grades being only related to the small amounts of W and V (Table 3.1). The amount of carbides in D3/D6 is higher than in D2, enabling better wear resistance but reduced toughness (compare Figures 5.14 and 5.18). But, the low amounts of Mo and V of D3 and D6 lead also to cost advantages in terms of alloy cost. Although limited in terms of hardenability and practically restricted to oil quenching, D3 and D6 grades have very simple heat treatment practices, which also serve for further cost reduction in metal-forming tools for blanking, cutting, and stamping. The reason is that the low amount of Mo and V does not require high hardening temperatures, and D3/D6 is usually hardened at 960°C (1760°F), while D2 is hardened at 1020°C (1870°F). The high amounts of carbides in D3 and D6 are also very effective in promoting high abrasive wear resistance, which is important for ceramic shaping tools and other tools subjected to pure abrasion.

Another important issue for D2 and D3 or D6 is the tempering treatment. For high hardness, ~60 HRC, these three grades are usually tempered at low temperatures of ~200°C (390°F). While this may lead to retained austenite effect (e.g., dimensional instabilities), this tempering conditions is very simple in terms of heat-treatment practices and furnace construction. For D3 and D6, the combination of simple hardening and tempering practices leads to very inexpensive heat treatment, which combined with the low alloy cost, makes those grades very cost effective and popular for low-demanding tools. For D2 tool steel, tempering at high temperature may also be applied, but the hardness is usually limited to the range 55–58 HRC (higher hardness is obtained only at higher hardening temperature, but problems of retained austenite and toughness may emerge from this practice, as explained in Section 5.3.1.3).

For all the high-Cr, high-C steels, the majority of carbides are of M_7Cr_3 type, which show high hardness, but depending on the amount and distribution, may also decrease the toughness. The amount of carbides, as explained in Section 5.3.2.2, is primarily determined by the solubility product of %C and %Cr (as seen in Figure 5.14). Therefore, in order to increase toughness, the 8%Cr steels were developed about 15–20 years ago, and they are very popular even today. As shown in Figure 5.18, those grades present a very good combination of strength, adhesive wear resistance, and toughness, which are the most important properties for cold work tooling that involve metal–metal working conditions, such as stamping, blanking, or punching. The lower amount of M_7C_3 carbides, which is the main reason for the improvement in toughness, would also negatively impact the wear resistance of those grades. However, there are factors that bring the wear resistance still to good levels, such as (1) the high hardness, up to 62 HRC after high-temperature tempering; (2) the presence of hard MC carbides, such as NbC; (3) the good carbide distribution, which is very important for adhesive wear conditions; and (4) elimination of retained austenite effects and improvement in toughness by high-temperature tempering (Figure 5.13), reducing microcracks and other wear mechanisms related to low toughness. Therefore, the adhesive wear behavior of 8%Cr steels tends to be better than that of D2, D3, and D6 grades, and the pure abrasive wear resistance is similar to that of D2 in many cases.

High-temperature tempering also improves the ability of 8%Cr steels to serve as a suitable substrate for surface treatments. The most common treatment to increase the surface hardness in tool steel is nitriding, and the most common coating (application of a fine layer of hard compound) is through PVD. Both can be applied to improve the surface wear characteristics, avoiding both adhesion and abrasion, without impairing the toughness of the core areas (i.e., the tool steel toughness). So, tool steel subjected to these treatments will have a stable microstructure during the surface treatment, which is done at ~400°C–550°C (750°F–1020°F), and a high hardness, especially in the case of PVD, avoiding penetration and cracking of the applied coating (egg-shell effect). By eliminating the retained austenite and promoting hardness to ~60–62 HRC after 530°C/550°C (990°F/1020°F) tempering, 8%Cr tool steels thus presents an ideal condition to the application of both nitriding and PVD coatings.

The disadvantages of 8%Cr steels are usually the higher cost of the alloy and of processing. All the rest of the properties described in the previous paragraph are provided by the presence of elements other than Cr, mainly Mo, V, and Nb, which leads to a cost increase due to the higher cost of these alloys. In addition, for obtaining the properties, 8%Cr steels have to be hardened and tempered at high temperatures, typically 1030°C and 530°C, respectively (1890°F and 990°F). There are actually possibilities to heat-treat 8%Cr steels at lower temperatures, but this is strongly not recommended: The toughness and wear resistance properties would be deeply impaired because the microstructural features described would be disrupted—no elimination of retained austenite and no formation of fine secondary hardening carbides in the tempered martensite structure. So, the change in heat treatment, especially compared to D3 and D6 grades, is a must when dealing with 8%Cr grades. Their applications are then often successful in highly demanding tools when the increase in performance pays for the cost increase, such as in fine blanking, stamping, or other sheet-forming processes.

Not usually a member of this class, the A2 and other A-series tool steels may also be compared with the traditional high-Cr, high C-steels of the D series. The lower Cr and lower C of these grades, when compared to D2, D3, or even 8%Cr steels, lead to very small formation of carbides, which increases the toughness but decreases substantially the wear resistance. Another characteristic of A-series grades is the high hardenability, the "A" referring to "air-hardening" steel. As with D2, A-series tool steels may have a high amount of retained austenite, and tempering at high temperatures is recommended to decrease its amount (but this also decreases the final hardness). The application of A2 tool steel is common for tools that are less demanding in wear but highly demanding in toughness, especially after high-temperature tempering. In addition, A2 steels may be applied in tools where oil-hardening tool steels, such as O1, show problems of hardenability. The major replacement of D-series steels by A2 is usually constrained by the lower amount of carbides and so the lower wear resistance of A2. Other grades in the A-series than A2 may present higher amount of carbides, such as A3 (1.3%C, 5%Cr, 1%Mo) or A7 (2%C, 5%Cr, 1%W, 4%V). But, those grades are less common today because they may be replaced by 8%Cr steels, in terms of cost benefit.

5.4.2 Low-Alloy Cold Work Steels: O-Series and W-Series

In this section, low-alloy cold work steels are considered, which are of low hardenability and thus hardenable only by oil (O-series) or water (W-series). There are some advantages, however, in those grades. O1 tool steel may be hardened in reasonable sizes, up to 50 mm (or 2 in.) (Figure 5.19). The composition with a high carbon content enables O1 tool steel to obtain high hardness after quenching (Figure 5.10), achieving ~60 HRC after low-temperature tempering (Figure 5.4); also, the low amount of alloy elements leads to very low cost for this grades. For larger tools and dies, the hardenability of O1 tool steel may not be enough. Also, the absence of carbides may compromise their wear resistance. Therefore, O1 and other steels in the O-series are often applied to small tools, especially those used for punching, coining, and powder pressing.

Still on the hardenability, W1 is much more limited, as shown in Figure 5.19, the hardened case being less than 6 mm (1/4 in.). While this limits its application to bigger tools, advantages are also seen in small tools: They may naturally form a hard and wear-resistant case, with a tougher and crack-resistant core. This combination is very helpful in punching and coning tools used, for example, in the production of screws and bolts. But today, with the advent of surface hardening technologies (carburizing, nitriding, and PVD coatings), the use of W-series steels is less common and they are rarely seen in tool steel distribution centers.

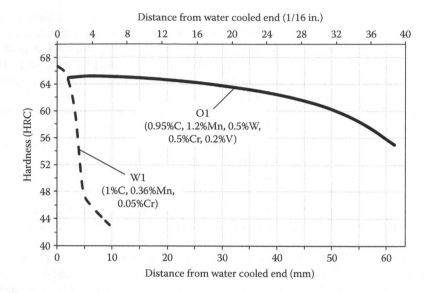

FIGURE 5.19 Typical hardenability curves for O1 and W1 steel. O1 was hardened at 830°C (1520°F) and the measurements refer to the Jominy end-quench test. W1 specimens are 19 mm cylinders, quenched in water. In comparative terms, the quenching is slower for O1, and if both steels were tested in the same method, the difference would be even higher. (Adapted from Roberts, G. et al., *Tool Steels*, 5th edn., ASM International, Materials Park, OH, 1998, pp. 133, 187.)

5.4.3 High-Speed Steels and Highly Alloyed PM Steels Used in Cold Work Tools

High-speed steels may be used in some cases for cold work tooling when the need for higher matrix hardness and hard carbides is necessary. The characteristics of those products will be shown later in this book (Chapter 7), but in general they may be understood as a refined version of cold work tool steels. In fact, high-Cr, high-C cold work steels were developed during the First World War as a possible substitute of high-speed steels [58]; the lack of high-temperature strength has made these grades today the D-series, restricted to cold forming, where they show enough high wear resistance.

Conversely, high-speed steels may be used in cold work tooling when higher hardness (65–68 HRC) than with typical cold work steels (60 HRC) is necessary, especially when this hardness has to be achieved after high-temperature tempering, in order to apply surface hardening treatments of nitriding or PVD. In addition, as seen in Table 5.2, the alloy carbides typically present in high-speed steels (MC and M_6C) show higher hardness than the Cr-rich M_7C_3 carbides of cold work steels. Therefore, when extreme conditions are reached in cold-form tooling, such as high-performance coated or nitrided tools, high-speed steels may substitute traditional cold work tool steels. Of course, the improvement in performance has to be sufficiently high to overcome the cost increase, because of the high alloy and heat-treatment costs of high-speed steels. Typically, M2, M35, and M42 may be applied under some conditions.

To combine all the aforementioned advantages (high hardness, high tempering resistance, and extremely hard carbides) with improvements in homogeneity mechanical properties, distortion, and toughness, high-alloy PM tool steels may also be applied in cold work tooling. Some of these PM grades are typically based on high-speed steels, such as M3:2 (1%C, 5%Mo, 6%W, 2%V), but others are based on high Nb or high grades, typically 4%–10%V. The addition of V is to promote higher amount of MC-type carbides and improvement of wear resistance. Basically, the homogeneity of properties has already been presented in this book, as shown in Figures 5.16 and 3.35. The specific combination of alloy elements for each will then balance the increase in wear resistance by changing the matrix hardness and the carbides (amount and type). And this balance is not standardized because the development of these grades is very dynamic. So, the results can be seen in publications from steel manufactures, such as in [4,5,59].

5.4.4 SHOCK-RESISTANT TOOL STEELS

In strict metallurgical concepts, shock-resistant grades, such as S1, would be considered very different than cold work steels. The chemical composition, heat treatment, and final properties make those steels actually closer to hot work grades. In fact, several modified hot work steels may be used in shock-resistant cold work tooling, such as high-C H13 or modified DIN 1.2714 or DIN 1.2721. The main concept behind those grades is the formation of a microstructure without carbides and with up to 55 HRC hardness, which in comparison to A-, D- or O-series tool steels causes a strong decrease in wear resistance, but at the same time promotes a strong increase in toughness (Figures 5.8 and 5.18).

Therefore, S1 and other typical hot work steels used in cold work tooling are applied in tools subjected to less abrasion, but with high demands on mechanical shocks and crack propagation. Examples are shear-cutting of thick plates and billets at room or moderate temperatures, in addition to chisel and other high-shocking components. In some cases, nitriding may be applied to increase the surface wear resistance, but the reduction of hardness after high-temperature tempering may restrict the mechanical strength and cause plastic deformation in some high-stress points.

REFERENCES

1. Uddeholm Tool Steel for Cold Work Tooling. Available at: http://www.uddeholm.com/files/. Accessed January 2016.
2. ÖNORM EN ISO 4957:1999. *Tool Steels*. Publisher and Printing: Österreichisches Normungsinstitut, 1020 Wien.
3. ASTM A681-08, Standard specification for tool steels alloy, ASTM International, West Conshohocken, PA, 2008.
4. Uddeholm Tooling. Technical catalogue for tool steels. Available online at: http://www.uddeholm.com/files/. Accessed December 2015.
5. Böhler Edelstahl. Technical catalogue for tool steels. Available online at: http://www.bohler-edelstahl.com/english/1853_ENG_HTML.php. Accessed December 2015.

6. Villares Metals. Technical catalogue for tool steels. Available online at: http://www. villaresmetals.com.br/pt/Produtos/Acos-Ferramenta. Accessed December 2015.

7. Annual Book of Standards. Standard terminology relating to wear and erosion. ASTM International, West Conshohocken, PA, vol. 03.02, 1987, pp. 243–250.

8. H. Brandis, E. Haberling, H.H. Weigard. Metallurgical aspects of carbides in high speed steels. *Processing and Properties of High Speed Tool Steels*, Eds., M.G.H. Wells, L.W. Lherbier, Warrendale, PA: TMS-AIME, 1980, pp. 1–18.

9. L.P. Tarasov. The microhardness of carbides in tool steels. *Metal Progress*, 54(6), 846, 1948.

10. P. Leckie-Ewing. A study of the microhardness of the major carbides in some high-speed steels. *Transactions of ASM*, 44, 348, 1952.

11. S. Karagös, Fischmeister, H.F. Niobium-alloyed high speed steel by powder metallurgy. *Metallurgical Transactions A*, 19A, 1395–1401, June 1988.

12. M.M. Khrushchov, M.A. Babichev. An investigation of the wear of metals and alloys by rubbing on adhesive surface. *Friction and Wear in Machinery*, American Society of Mechanical Engineers, ASME, 12, 1–12, 1956.

13. K.-H. Zum Gahr. *Microstructure and Wear of Materials*. New York: Elsevier, 1987.

14. R.A. Mesquita, C.A. Barbosa, Failure analysis in tool steels. *Failure Analysis of Heat Treated Steel Components*, L.C.F. Canale, R.A. Mesquita, G.E. Totten (Eds.). Materials Park, OH: ASM International, 2008, Chapter 11, pp. 311–350.

15. R.A. Mesquita, C.A. Barbosa. Powder metallurgy and spray formed 10%V cold work tool steels, a comparison. *Proceedings of Euro PM 2004 (Powder Metallurgy World Congress & Exhibition)*, 5, 53–64, 2004.

16. I. Schruff, V. Schüler, C. Spiegelhauer. Advanced tool steels produced via spray forming. *Proceedings of the Sixth International Tooling Conference—The Use of Tool Steels: Experience and Research*, Eds. J. Bergstrom, G. Fredriksson, M. Johansson, O. Kotik, F. Thuvander, Karlstad, Sweden, 2002, pp. 1159–1179.

17. D.H. Buckley. *Surface Effects in Adhesion, Friction, Wear and Lubrication*, New York: Elsevier, 1981, pp. 315–427.

18. K.C. Ludema. Sliding and adhesive wear. *Friction Lubrication and Wear Technology*, ASM Handbook Materials Park, OH: ASM International, vol. 18, pp. 236–241, 1992.

19. Uddeholm Tooling. Technical catalogue on vancron tool steel. Available at: www. uddeholm.com/files/PB_vancron_40_english.pdf. Accessed in March 2015.

20. R.A. Mesquita, C.A. Barbosa. An evaluation of wear and toughness in cold work tool steels. *Tecnologia em Metalurgia e Materiais*, 2, 12–18, 2005. (In Portuguese, original title: *"Uma avaliação das propriedades de desgaste e tenacidade em aços para trabalho a frio"*).

21. R.M. Hemphill, D.E. Wert. Impact and fracture toughness testing of common grades of tool steels. *Tool Materials for Molds and Dies*, Eds. G. Krauss, H. Nordberg, Golden, MO: Colorado School of Mines Press, 1987, pp. 66–69.

22. J.W. Sullivan. Preventing failures in cold forming tools. *ASM Metal Engineering Quarterly*, 13, 31–41, February 1973.

23. J.R. Davis (Ed.). *Tool Materials*, ASM Specialty Handbook. Materials Park, OH: ASM International, 1995, pp. 421–441.

24. G. Krauss. Martensitic transformation, structure and properties in hardenable steels. *Hardenability Concepts with Application to Steel*, D.V. Doane and J.S. Kirkaldy, eds., AIME, Warrendale, PA, 1978, pp. 229–248.

25. J.R. Davis (Ed.). *Tool Materials*, ASM Specialty Handbook. Materials Park, OH: ASM International, 1995.

26. Uddeholm Tooling. Grinding of tool steel, Technical brochure. Available at: www. uddeholm.com/files/TB_grinding-english.pdf. Accessed March, 2016.

27. ASM Handbook. *Failure Analysis and Prevention*, 8th edn. Materials Park, OH: American Society for Metals, vol. 10, 1975, pp. 500–507.
28. M. Cohen. Retained austenite. *Transactions of ASM*, 41, 35–94, 1949.
29. B.L. Averbach, S.A. Kulin, M. Cohen. The effect of plastic deformation on solid reactions, Part II: The effect of applied stress on the martensite reaction, *Cold Working of Metals*, American Society for Metals, Cleveland, OH, 1949.
30. O. Zmeskal, M. Cohen. The tempering of two high carbon-high chromium steels. *Transactions of ASM*, 31, 380, 1943.
31. Uddeholm Tooling. Product Brochure Uddeholm Sverker 21. Available online at: http://www.uddeholm.com/files/PB_Uddeholm_sverker_21_english.pdf. Accessed February 2016.
32. Uddeholm Tooling. Product Brochure Uddeholm Sleipner. Available online at: http://www.uddeholm.com/files/PB_Uddeholm_sleipner_english.pdf. Accessed February 2016.
33. R.S.E. Schneider, R.A. Mesquita. Advances in tool steels and their heat treatment part 1: Cold work tool steels. *International Heat Treatment & Surface Engineering*, 4, 138–144, 2010.
34. F. Kayser, M. Cohen. Carbides in high speed steel—Their nature and quantity. *Metal Progress*, 61(6), 79, 1952.
35. R.A. Mesquita, C.A. Barbosa. Hard alloys with lean composition. Patent: US 8,168,009), 2006.
36. R.A. Mesquita, C.A. Barbosa. High-speed steel for saw blades. Patent: US2009/0123322, 2006.
37. E. Rudy, F. Benesovsky, L. Toth, Untersuchung der Dreistoffsysteme der Va- und VIa-Metalle mit Bor und Kohlenstoff, *Zeitschrift für Metallkunde*, 54, 345, 1963. (JCPDS record # 38–1364).
38. J. Hanawalt, H. Rinn, L. Frevel. Chemical analysis by X-ray diffraction, *Analytical Chemistry*, 10, 457, 1938. (JCPDS record # 01-1159).
39. United States National Bureau of Standards. *Monography* 25, 21, 95, 1984 (JCPDS record # 35-0787).
40. J. J. Zhu, J. Jiang, C.J.H. Jacobsenb, X. P. Linc. Preparation of Fe–Mo–C ternarycarbide by mechanical alloying. *J. Mater. Chem.*, 11, 864–868, 2011.
41. H.J. Goldschmidt. Interplanar spacings of carbides in steels. *Metallurgia*, 40, 103, 1949. (JCPDS record # 05-0720).
42. R. Benz, J.F. Elliott, J. Chipman. Thermodynamics of the carbides in the system Fe-Cr-C. *Metallurgical Transactions*, 5, 2235–2240, 1974.
43. E. Ohman. X-ray investigations on the crystal structure of hardened steel. *Journal Iron and Steel Institute*, 123, 445, 1931.
44. E. Elsen, G. Elsen, M. Markworth, Zum Stand der mit Vanadium hochlegierten Schnellarbeitsstähle. *Metall*, 19, 334–345, 1965.
45. G.B. Brook, J.M.G. Crompton. Fulmer Report R 319/4, Fulmer Research Institute, 1971. Apud: S. Karagoz, H. Fischmeister, *Metallurgical Transactions A*, 19Z, 1935–1941, 1988, 0.
46. S.R. Keown, E. Kudielka, E. Heisterkamp. Replacement of vanadium by niobium in S6-5-2 high-speed tool steels. *Metals Technology*, 7, 50–57, 1980.
47. E. Haberling, P. Giimpel. *Thyssen Edelstahl Technische Berichte*, 6, 127–131, 1980.
48. K. Yamamoto, S. Inthidech, N. Sasaguri, Y. Matsubara. Influence of Mo and W on high temperature hardness of M7C3 carbide in high chromium white cast iron. *Materials Transactions*, 55(4), 684–689, 2014.
49. L.P. Tarasov. The microhardness of carbides in tool steels. *Metal Progress*, 54(6), 846–847, 1948.

50. Reference charts for evaluating the carbide distribution in high carbon 12%Cr steels, printed in 1991. *Material Purchase Specification (Zukaufspezifikation Material)* B.hler Edelstahl, Kapfenberg, Austria. May, 2004.

51. P. Rayand, P.K. Pal. High speed steel semis—Alternate production routes. *World Conference on Powder Metallurgy*, London, U.K.: Institute of Metals, vol. 2, 1990, pp. 159–168.

52. P. Hellman. As-HIPed APM high speed steels, *Metal Powder Report*, 47(6), 25, 1992.

53. R. Schneider, A. Schulz, C. Bertrand, A. Kulmburg, A. Oldewurte, V. Uhlenwinkel, D. Viale. The performance of spray-formed tool steels in comparison to conventional route material. *Proceedings of the Sixth International Tooling Conference—The Use of Tool Steels: Experience and Research*, Eds. J. Bergstrom, G. Fredriksson, M. Johansson, O. Kotik, F. Thuvander, Karlstad, Sweden, 2002, pp. 1111–1124.

54. J.B. Forrest, R.R. Pratt, J.S. Combs. Production of spray deposits. US Patent 5,472,038, Osprey Metals Ltd., May 12, 1995.

55. C. Spiegelhauer. Industrial production of tool steels using the spray forming technology. *Proceedings of the Sixth International Tooling Conference—The Use of Tool Steels: Experience and Research*, Eds. J. Bergstrom, G. Fredriksson, M. Johansson, O. Kotik, F. Thuvander, Karlstad, Sweden, 2002, pp. 1101–1109.

56. A. Mendanha, H. Goldenstein, C.E. Pinedo. Toughness of D2 cold work tool steel— Part I: The influence of the initial microstructure. *Proceeding of the Encontro de Integrantes da Cadeia Produtiva de Ferramentas, Moldes e Matrizes*, São Paulo, Brazil: ABM, October 2003. (In Portuguese, original title: tenacidade do aço ferramenta para trabalho a frio AISI D2—parte I influência da microestrutura de partida).

57. G. Roberts, G. Krauss, R. Kennedy. *Tool Steels*, 5th edn. Materials Park, OH: ASM International, 1998, pp. 133, 187.

58. P. Jurči. Cr-V ledeburitic cold-work tool steels. *Materiali in Tehnologije/Materials and Technology*, 45(5), 383–394, 2011.

59. Bohler. High performance steels produced by powder metallurgy methods. Available at: http://www.bohler.at/deutsch/files/downloads/ST035DE_Microclean_2013.pdf. Accessed March, 2016.

6 Plastic Mold Tool Steels

6.1 INTRODUCTION TO PLASTIC MOLD TOOLING

Plastic mold steels constituted the smallest group within the different types of tool steels before the 1980s. Because of the tremendous increase in the plastic industry, the mold steels are much more important in the tooling industry. Today, the mold steels are the largest group on a tonnage basis. The term "plastics" in fact refers to thermoplastics, which means the group of polymers is able to be formed after heated. All plastics produced today require at least one mold in their manufacturing process, in which the forming operation will be performed. As a consequence, the growth of the plastic industry will directly affect the consumption of materials for molds, in particular plastic mold tool steels, which are today an essential part of the plastic industry chain.

Within the whole plastic industry, several operations may be used to produce a plastic part, such as injection molding, blow molding, and extrusion, as well as a combination of processes such as injection and extrusion (for bottles) or extrusion and blowing (for plastic films). In a volume basis, however, most tool steels are used in injection molding. This process is described in some detail in Figure 6.1, and examples of plastic molds are shown in Figure 6.2. The process basically involves the high-pressure injection of liquid plastic into the tool steel cavity, similar to aluminum die-casting process explained in Chapter 4 (see Figure 4.14). The differences are mainly in the temperature requirement—which is usually below 200°C and thus does not require heat or wear resistance from the tool steel—and the surface quality requirement—which is extremely high in the plastic industry and less important in aluminum die-casting.

Since the temperature and wear requirements in most injection molding processes are low, plastic mold steels tend to have leaner alloy designs and metallurgical properties when compared to other tool steels. In fact, they may be considered simple or easy-to-produce steels within steel mill processes. However, the requirements for the final applications, in terms of surface finishing, and the involved cost of mold manufacturing make plastic mold steels tricky. To understand this point, Figure 6.3a may be considered, which shows the main cost item of a typical mold. By having an unusually low alloy content, the tool steel cost is approximately 10%–20% of the total mold cost. The remaining 80% of the cost is incurred during the manufacturing steps, which comprise the amount of machining added to the molds, as shown in the examples of Figure 6.3b.

In spite of the low significance of the tool steel in the final cost, it is important to consider that all the manufacturing operations are directly affected by the plastic mold steel used and its quality. For example, a tool steel with heterogeneous hardness or high amount of inclusions may require much longer machining, not only increasing the manufacturing cost but also compromising the final surface quality and the performance of the final mold. Because of that, the mold steel cost is often compared to an iceberg, where only the minor part of the cost is seen

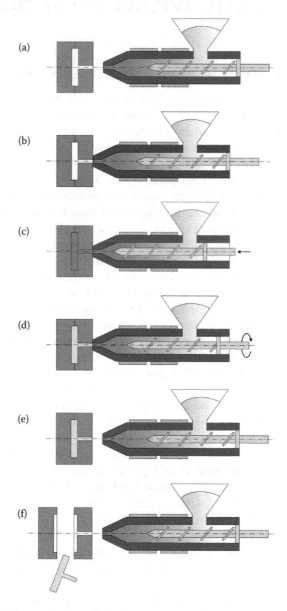

FIGURE 6.1 Diagramatic view of plastic injection molding, showing the main operations. (a) Close mould, (b) apply nozzle, (c) inject plastic, and hold pressure, (d) allow part to set, for next part meter compound and plasticize, (e) remove nozzle, and (f) open mould and eject part. (From Hippenstiel, F. et al., *Handbook of Plastic Mould Steels*, Edelstahlwerke Buderus AG, Wetzlar, Germany, 2001, pp. 17, 40, 67, 125–149, 248–251.)

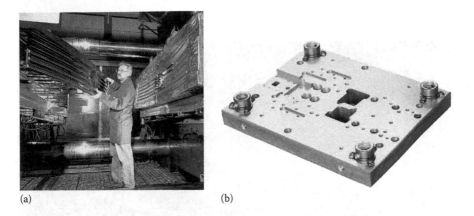

(a)   (b)

FIGURE 6.2 Images showing (a) a large injection mold, after extraction of an injected front frame for a truck, and (b) mold plates used as standard element for manufacturing small molds. (From Hippenstiel, F. et al., *Handbook of Plastic Mould Steels*, Edelstahlwerke Buderus AG, Wetzlar, Germany, 2001, pp. 17, 40, 67, 125–149, 248–251.)

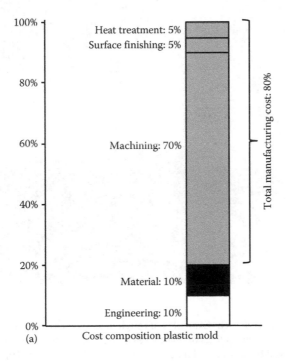

FIGURE 6.3 (a) Cost composition of a typical injection, based on common industrial data for plastic mold manufacturing. (From Hippenstiel, F. et al., *Handbook of Plastic Mould Steels*, Edelstahlwerke Buderus AG, Wetzlar, Germany, 2001, pp. 17, 40, 67, 125–149, 248–251.)   (*Continued*)

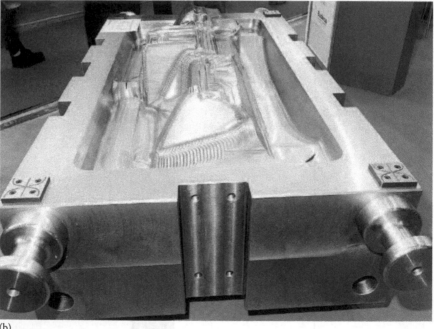

(b)

FIGURE 6.3 (Continued)   (b) Examples of rough machined cavities for car bumpers, with weight of more than 10 tons. Observe that those cavities were massive blocks, from which all volume was removed by sawing and machining (rough milling). (From Hippenstiel, F. et al., *Handbook of Plastic Mould Steels*, Edelstahlwerke Buderus AG, Wetzlar, Germany, 2001, pp. 17, 40, 67, 125–149, 248–251.)

(for example the tool steel cost) while the majority is based on the manufacturing-related processes. The quality of the mold steel, therefore, affects the overall cost. By investing in a better quality and higher priced steel, the cost of operations may then be reduced, and then reduce the total mold cost.

As discussed previously, the main aspects of mold steels are (1) the interaction to the manufacturing processes, especially machining and surface finishing, and (2) the large sizes and the achieved hardness through the section size and (3) high surface requirements. In terms of metallurgical properties, such as strength, wear resistance, and hot hardness, the mold steels tend not to have high requirements and thus have simpler alloy designs. With those aspects in mind, the present chapter was prepared, discussing first the hardening characteristics of large mold steel blocks and the importance of volume aspects in terms of segregation (Section 6.2). Then, the main interrelationships between the mold steel and the surface finishing are presented (Section 6.3).

The focus of this chapter are the most common plastic mold steels, with their chemical composition shown in Table 6.1. There are always exceptions, of course, of special molds for abrasive plastics (such as fiber- or particle-reinforced plastics) or molds for extremely high polishability, for optical applications. But, those special cases form a minority and, on a volume basis, more than 90% of the molds are made of steel grades shown in Table 6.1 or variations on those grades.

In fact, most molds are made of a few steel grades. The majority are the modified ASTM P20 grades, well known by their DIN number. This includes, in the order of importance, DIN 1.2738, followed by 1.2311 and in smaller amount by 1.2312 [1]. All focus on 32 HRC hardness and are referred to here as the modified P20 or P20 type grades. In addition, 40 HRC grades are used, although to a lesser extent. DIN 1.2711 is an important representative of this family, and is very similar to the DIN 1.2714, discussed within the hot work steels (Chapter 4).

TABLE 6.1

Chemical Composition of the Main Plastic Mold Steels, Showing the Commonly Observed Target Values

Plastic Mold Tool Steels			Most Common Chemical Composition (wt.%)							
ASTM	EN/DIN	UNS	C	Si	Mn	Cr	V	W	Mo	Other
P20	—	T51620	0.35	0.4	0.5	1.7	—	—	0.4	—
P20 Mod	1.2738	—	0.38	0.3	1.5	2.0	—	—	0.2	Ni = 1.0
P20 Mod	1.2311	—	0.38	0.3	1.5	2.0	—	—	0.2	—
P20 Mod	1.2312	—	0.38	0.3	1.5	2.0	—	—	0.2	S = 0.070
6F2 Mod	1.2711		0.52	0.2	0.7	0.8	0.1	—	0.3	Ni = 1.7
420	1.2083	S42000	0.42	0.4	0.4	13.0	—	—	—	—
422 Mod	1.2316	—	0.36	0.3	0.7	16.0	—	—	1.1	Ni = 0.80

Note: The specific limits and ranges are given by the standards ASTM A681 and Euro Norm ISO 4957 (old DIN 17350 and EN 10027), according to References 2 and 3. In this chapter, the DIN numbers are mostly used as reference for the plastic mold steels.

6.2 STRENGTH, HARDNESS, AND SECTION SIZE
IN PLASTIC MOLD STEEL BLOCKS

As in other types of tool steels, mechanical strength is the most important characteristic of plastic mold steels. This is achieved by heat treatment, usually by hardening and tempering. In practical terms, the strength is measured and controlled by the hardness. In this chapter, hardness will be discussed extensively. The most significant aspect of hardness is withstanding the closing forces during the injection molding process. This is especially important in molds with thin walls, designed to improve heat exchange while maintaing high closing forces, which actually avoids the plastic from leaking. As will be shown in Section 6.3.1, the mold steel hardness is also important to the finishing process, especially the final polishing quality. So, mold designers usually define the minimum hardness as the initial parameter of a mold steel.

The hardness for the main plastic mold steels, after hardening and tempering, can be seen in the graphs shown in Figure 6.4. In order to reduce the manufacturing complexity, in terms of mold logistics (never easy for pieces with more than 10 tons), many molds are commonly found in the market in the prehardened condition. As opposed to hot work or cold work tool steels, this means many plastic mold steels are hardened and tempered by the tool steel producer, with large furnaces and large quenching tanks, usually with oil quenching. For example, mold steels such as DIN 1.2738, 1.2311, 1.2312, and 1.2711 are almost exclusively found for purchase in this condition. Because the wear characteristics are usually low, the hardness of plastic mold steels is often in the range of 30–40 HRC, which is also another reason for their availability in the prehardened condition. So, the tempering temperatures are often in the range of 500°C–600°C (Figure 6.4).

One important remark has to be made here in terms of the size effect for mold steel blocks. The heat-treatment response shown in Figure 6.4 is for samples of 25 mm (1 in.) size, which is clearly different from large tool steel blocks, weighing often more than 10 tons. This size aspect becomes even more critical when considering that, for cost purposes, the chemical composition of the common plastic mold steels (P20 variations) is very lean in terms of alloying elements, which are the ones important for hardenability (thru hardness). So the core regions of a plastic mold may show lower hardness and then not respond adequately to the mold operation (closing forces) and to the finishing aspects (polishing or texturing). In this respect, Figure 6.5a compares the core hardness of three steel grades, showing that for 1.2311 the hardness decreases in the core areas of large blocks.

The metallurgical reason for all those differences lies on the transformation upon cooling, especially by the formation of pearlite. In the heat-treatment literature, martensite is always treated as the aim phase when quenching steel, but it can hardly ever be formed in large blocks, especially in low-alloy-containing steels such as P20-type mold steels. Those large blocks usually have bainite as the main constituent, which is not detrimental, because bainite also shows reasonably high hardness, tempering resistance, and homogeneity, because of the fine dispersion of cementite particles (of nanometric size) in a needle-like matrix. There is a difference in hardness between tempered bainite and tempered martensite, though, which is

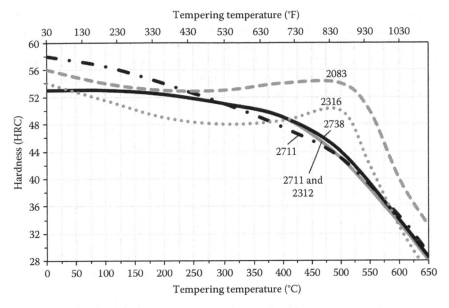

ASTM	DIN	Hardening	Hardness (HRC)								
				Tempering, 2 × 2 h							
Plastic tool steels			As quench	200°C (400°F)	300°C (570°F)	400°C (750°F)	500°C (930°F)	550°C (1020°F)	600°C (1110°F)	650°C (1200°F)
P20 Mod	1.2738	880°C (1610°F)	53	52	51	49	44	40	34	28
P20 Mod	1.2312	880°C (1610°F)	53	52	51	49	43	39	34	28
6F2 Mod	1.2711	880°C (1610°F)	58	54	51	48	44	40	36	29
420	1.2083	1020°C (1870°F)	56	53	53	54	54	50	40	33
422 Mod	1.2316	1020°C (1870°F)	54	49	48	49	50	42	33	26

FIGURE 6.4 Tempering diagrams for the most common plastic mold tool steels, showing hardness after two times tempering for 2 h. Data from brochures of traditional tool steel producers, References 1, 4, and 6, for samples of 25 mm (1 in.) quenched in oil. Observe that the plastic molds are usually made of large blocks, where the as-quenched microstructure of low alloy grades (mod P20) is usually bainitic and then the hardness after tempering is lower than in the tempering diagrams. A rule of thumb is considering the hardness of large molds with 4 HRC lower temper than the graph above, provided that only bainite and no significant amounts of pearlite are formed in the microstructure.

usually between 2 and 4 HRC depending on the tempering conditions. (See remark in Figure 6.4). However, when considerable amounts of pearlite are formed, the decrease in hardness is much stronger. The reason is that the carbide in pearlite will be forming lamellar cementite, and the remaining microstructure will be ferrite, which has very low hardness.

This relationship between core hardness and pearlite may be easily seen in the continuous transformation diagrams (CCT) of DIN 1.2738, 1.2311, and a modified 1.2738. The cooling curves that cross the pearlite field show a strong decrease in hardness (see the HV values in the diagrams), while the hardness of bainite is much higher. The decrease in hardenability shown in Figure 6.5a, leading to low hardness in large bars of 2311, is then easily explained by the pearlite formation in short times for this steel. In other words, the core regions always cool down more slowly than the surface and may form pearlite if the hardenability of the steel does not "push the pearlite field to sufficiently long times." On the other hand, the changes in chemical composition (from DIN 2311 to DIN 2738, and from DIN 2738 to the 2738mod) continuously push the pearlite field to the right, meaning less amount of pearlite in slow cooling times and, as an overall consequence, higher hardness in core regions of large bars.

The above discussion on the size limitations is important when considering the practice of cutting small pieces from master blocks. Large molds, with 10 tons or more, will always require large tool steel blocks, often with dimensions exceeding 1000 mm section size (40 in.). However smaller molds can be made from smaller forged blocks, especially for dimensions bellow 500 mm. The strategy of not cutting small blocks from large master blocks is interesting because it guarantees that the hardness will be more homogenous. This is due to the absence of pearlite, and the presence of uniform bainite which is easier to be obtained in quenching of smaller blocks. Another factor, that was not yet mentioned, is the possibility for macrosegregation in large blocks. The presence of microsegregation is natural from solute distribution during solidification and, as explained in Chapters 3 and 4, occurs in any kind of conventionally cast alloys and has extension of hundreds of microns. Macrosegregation, however, is mostly related to large cast pieces, such as large tool steel ingots, especially when weighting more than 10 tons. Figure 6.6 shows this effect of increasing the ingot size and increasing macrosegregation.

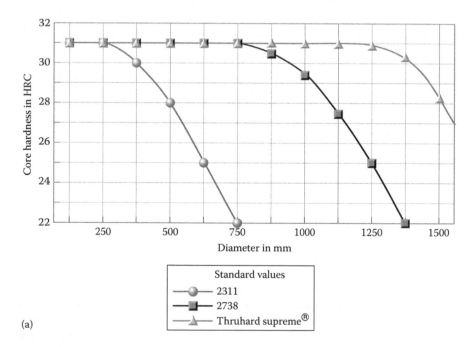

(a)

FIGURE 6.5   The challenges in large molds. (a) Hardness and section size for modified P20 grades, all with surface hardness of 32 HRC.   (*Continued*)

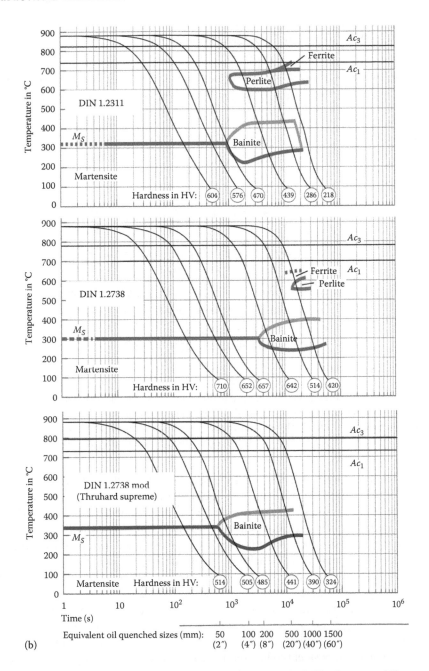

FIGURE 6.5 (*Continued*) The challenges in large molds. (b) Continuous cooling transformation (CCT) diagram for 1.2738 and 1.2711 steels, showing the occurrence of pearlite in earlier times for the 1.2711 steel, which explains the decrease of core hardness in the large-diameter blocks. Thruhard Supreme is a developed composition by Buderus Edelstahl. (Results adapted from Hippenstiel, F. et al., *Handbook of Plastic Mould Steels*, Edelstahlwerke Buderus AG, Wetzlar, Germany, 2001, pp. 17, 40, 67, 125–149, 248–251.).

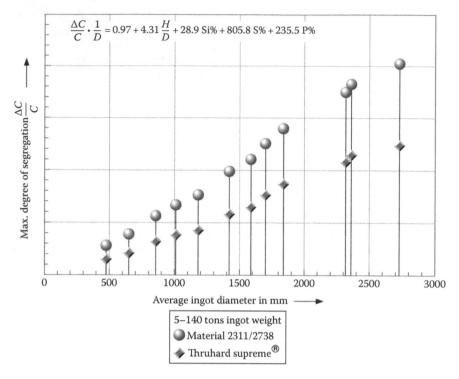

$$\frac{\Delta C}{C} \cdot \frac{1}{D} = 0.97 + 4.31 \frac{H}{D} + 28.9 \, Si\% + 805.8 \, S\% + 235.5 \, P\%$$

5–140 tons ingot weight
● Material 2311/2738
◆ Thruhard supreme®

FIGURE 6.6 Measurement of macrosegregation versus the size of the ingot. The degree of segregation is given by evaluating the local carbon content in relation to the nominal carbon in the steel composition ($\Delta C/C$). (From Hippenstiel, F. et al., *Handbook of Plastic Mould Steels*, Edelstahlwerke Buderus AG, Wetzlar, Germany, 2001, pp. 17, 40, 67, 125–149, 248–251.)

The theory is well described in Reference 7, but, in short, macrosegregation deals with the movement of segregated liquid, enriched in carbon and other alloying elements within the steel ingot during the solidification. After the solidification is completed, chemical differences may be found within several millimeters range with enriched areas in the core of hot-top areas of the ingot. Those areas will present different behaviors of hardenability, mechanical properties, and also response to machining, polishing, or texturing. So macrosegregation has to be avoided and should always be tolerated when there is a clear need for a large piece. In other words, the use of large blocks to produce small molds has to be done with caution to avoid macrosegregation problems, in addition to the hardness variation.

Before ending this section, it is also important to mention that there are processing parameters in the tool steel production that can reduce and intensify macrosegregation: notably the casting conditions, the ingot geometry, and the chemical composition. Actually, Figure 6.6 also presents a new grade that, due to changes in chemical composition, is less prone to macrosegregation. Nevertheless, macrosegregation will always exist in large blocks because they originate in large ingots with very long solidification times. So the subsequent processes, such as heat treatment, machining, polishing, and mainly texturing, have to be adapted when dealing with large mold steel blocks. In practice, it is often said that "the steel is acting different" during the production process. While this is true, it is also important to remember that large blocks, which come from large ingots, will always be less homogenous and then will show different characteristics during mold manufacture. And the reason may not the lack of quality of the steel but the segregation effects, which are normal in large ingots, dictated by nature and so with limitations for minimization.

6.3 SURFACE QUALITY AND THE TOOL STEEL METALLURGICAL CHARACTERISTICS

In particular to the injection molding process, molds play an essential role in the final quality of the produced part, especially with regard to surface quality [1]. Because of their low viscosity and surface tension, molten plastics are able to reproduce practically all visible details on the mold surface. While this is beneficial to the production of different surface appearances on plastic parts, defects on the mold may also be easily shown on the produced parts. For this reason, the surface of plastic molds are often carefully polished to obtain a shinny appearance of the produced parts.

In other situations, details may be intentionally introduced on the mold surface, in order to be reproduced on the injected part. This may include matted surfaces or specially designed appearances. Those surfaces are named textures* and a second important aspect in mold steel quality is their interaction or response to the texturing processes. Textures are typically applied to automotive parts for the interior of vehicles or in plastic parts for home use in order to change the visual appearance or improve friction and the handling of the part. Figure 6.7 shows examples of polished and textured plastics. The next sections deal with the metallurgical aspects important for polishing and texturing responses (or etchability).

6.3.1 POLISHING BEHAVIOR OF PLASTIC MOLD STEELS

The production of polished parts is critical for injection-molded parts for the purpose of showing a refined and shiny appearance or for optical purposes, such as headlight lens for vehicles, cell phone parts, household appliances and several other plastic products. Several metallurgical aspects of the tool steel interact during mold polishing. The main parameters are the tool steel hardness and the amount and distribution of inclusions, as discussed in the following paragraphs.

In terms of *inclusions*, they may be defined as phases found in the microstructure of all steels and metallic alloys. More specifically, they are oxides, sulfides, and other complex ionic or covalent phases found in the steel microstructure due to the reactivity of metallic atoms with oxygen or sulfur. All those particles are often simply referred to as *inclusions* (instead of the more appropriate technical word "nonmetallic inclusions") and are detrimental to mechanical properties, such as toughness and fatigue. Inclusions also negatively affects the polishing response, also called *polishability*, of tool steels. During polishing, undissolved carbides also may

---

* Graining, photo-etching or just etching may be used to refer to process to modify a plastic mold surface. This book, however, uses only the word texturing to refer to this aspect.

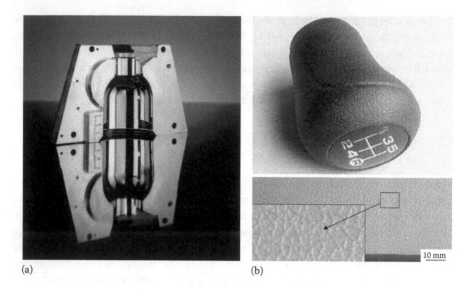

(a)   (b)

FIGURE 6.7 Example of surface finishing in plastic molds. (a) Injection mold for a coffee pot, with a mirror-like surface finish. (b) Textured part (image above) and surface of the textured mold (image below). (From Hippenstiel, F. et al., *Handbook of Plastic Mould Steels*, Edelstahlwerke Buderus AG, Wetzlar, Germany, 2001, pp. 17, 40, 67, 125–149, 248–251, with the textured surface included by the present book author.)

act like nonmetallic inclusions. For this reason, plastic mold steels usually do not contain undissolved carbides.

The effect of inclusions to the polishing process may be understood by analyzing Figure 6.8. During the polishing process, nonmetallic inclusions are extracted and small holes are formed. Before and after particle removal, larger areas may be affected, thus leading to polishing defects called "pin-holes" [8]. As a consequence, greater number and size of inclusions lead to lower polishability regardless of the inclusion type.

The immediate conclusion from this discussion is that mold steels should have very low levels of inclusions or, in other words, high microstructural cleanliness. While this is true, there are industrial process and cost limitations for the removal of inclusions due to the natural reaction of molten steel with oxygen and other impurities. Nonetheless, many metallurgical processes used in the steel manufacturing reduce inclusions to very small amounts; a small review on these methods is given in Section 6.3.4 and also detailed in Chapter 2.

In addition to the microstructural cleanliness (absence of inclusions), *mold steel hardness* is also very important for polishability (see also reference [8]). Polishing is a process of small abrasion, in which the polishing medium (diamond, alumina, etc.) is constantly scratching the surface. When the hardness increases, the penetration of the polishing particles inside the steel decreases (similar to what was shown in Figure 3.23). This leads to smaller micro-scratches, resulting in a mirror-like appearance to the naked eye. The quality after polishing will then depend directly on the hardness: the harder the steel, the better the polished appearance (Figure 6.9a).

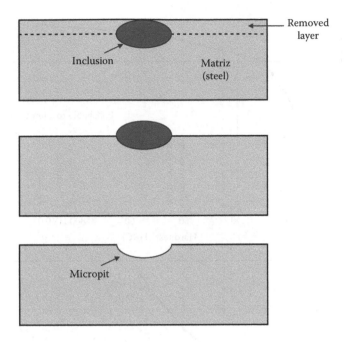

FIGURE 6.8 Pin-hole defects caused by the removal of inclusions during polishing. (Modified from Mesquita, R.A. and Schneider, R., *Exacta, São Paulo*, 8(3), 307, 2010.)

Another important result of increasing the steel hardness is the minimization of over-polishing problems, also known as the "orange-peel" phenomenon. This is caused by localized strain hardening on the mold surface by the subsequent deformation induced on the surface. After the start of over-polishing effect, surface roughness does not decrease with polishing time, but rather increases with the continuation of the polishing process (see the behavior of the curves in Figure 6.9b). The solution in such cases is regrinding, to remove the damaged area, and repeating the polishing process. When this is not possible, the over-polishing effect cannot be removed, and the quality of the injected plastic part is impaired. By increasing the hardness, the accumulation of strain at the surface is less likely, and thus the possibility of overpolishing is decreased. An example in Figure 6.9b shows the effect of doubling the hardness on the damage by overpolishing. Two other effects may accelerate overpolishing. First, macro surface defects, such as machining, grinding, sanding, or electrodischarge machining (EDM) defects. These increase the points of local stress concentration (stress riser effect) and thus enable faster development of surface strain hardening. Second, microstructure inhomogeneity points, such as decarburization or excessive segregation, cause nonconstant hardness and may accelerate the development of strain hardening.

To summarize the tool steel characteristics, two correlated features determine the polishability: (1) the steel cleanliness in relation to nonmetallic inclusions, and (2) hardness and microstructural homogeneity (microhardness homogeneity). And,

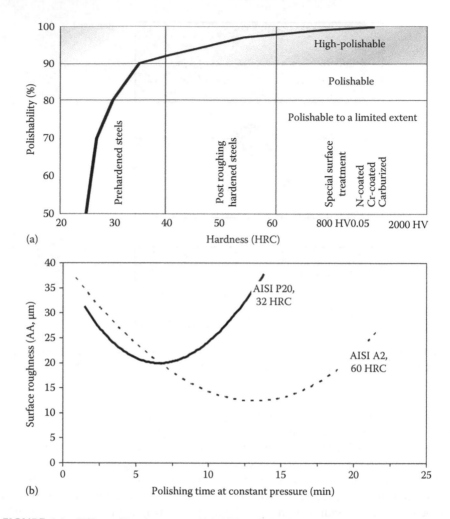

FIGURE 6.9 Effect of hardness on polishability of plastic molds. (a) Surface quality after polishing. (From Hippenstiel, F. et al., *Handbook of Plastic Mould Steels*, Edelstahlwerke Buderus AG, Wetzlar, Germany, 2001, pp. 17, 40, 67, 125–149, 248–251.) (b) Surface roughness and the over-polishing phenomenon—increase of roughness with the increase in polishing time. (From Uddeholm Technical Datasheet, Polishing mould steel, Available at: http://www.uddeholm.com/files/polishing-english.pdf, Accessed in March 2016.)

as explained in the previous paragraph, the final polishing quality also strongly depends on macro or micro defects, the first related to production process of the mold, and second to the tool steel microstructure.

6.3.2 Correlation between Polishability and Machinability

It is well known that the metallurgical parameters affecting polishability also influence other mechanical properties and the manufacturing-related properties of the tool steels. This is especially relevant for the machining behavior in metal-cutting processes, in most cases in an inverse way. And the opposite is critical, because while polishing determines the quality of the final product and the cost of finishing steps, machining constitutes most of the cost to convert a block of tool steel into a finished plastic mold, involving many hours of cavity and hole drilling and machining.

Going through the main parameters, the effect of inclusions in machining behavior of tool steels, or machinability, is first considered. Hard inclusions, such as oxides (and undissolved carbides), negatively affect machinability, by creating hard obstacles for chip removal during machining, thus causing abrasive wear on machining tools. In this respect, machinability and polishability follow the same trend. However, there are also soft inclusions, such as MnS, that improve machinability by facilitating chip-breaking and lubrication in the shearing zone of the chip. Consequently, high-S, high-machinability mold steels show an opposite behavior in terms of polishing and machining. This is the case of DIN 1.2312 mold steel, which contains a large amount of MnS inclusions. This mold steel enables tremendous improvement in cutting parameters during machining, but causes problems during the polishing process. For instance, this steel grade may be used in parts with high machining content but where the surfaces will not be polished, such as the male components of the mold, which get in touch with only the back (not appearing) regions of the produced parts.

While the increase in hardness improves the quality of polished molds, it also affects the cutting forces and the wear of cutting tools used in machining. The harder the steel, the lower the machining parameters. More specifically, an increase in hardness leads to an increase in strength, which causes higher stresses in the machining shearing zone, which reduces the tool life. This limits the machining time and consequently decreases machinability. So, again, the conditions for good polishing (high hardness) are opposite to easy machining (low hardness).

The correlation of these four characteristics (cleanliness, hardness, polishability, and machinability) is displayed in Figure 6.10, which clearly illustrates the compromise between machinability and polishability, both as a function of hardness. The beneficial effect of MnS inclusions is illustrated in Figure 6.10b, comparing the machinability index for a given hardness, which is increased when the S content increases. It is also interesting to compare the 40-HRC precipitation hardening steels which are refined by vacuum arc remelting (VAR) and thus may combine high S with sufficient polishability. Details on electro-slag remelting (ESR) and VAR refining processes are given in Section 6.3.4.

6.3.3 Texturing of Plastic Mold Steels

Textured or grained surfaces are produced in parts through plastic injection molding when the mold surface is designed to reproduce different patterns, such as leather-like matted surfaces or practically any kind of surface design. There are many reasons for texturing, but it is generally applied to improve the appearance of a plastic part or change the surface-friction coefficient, making it easier to handle a given part.

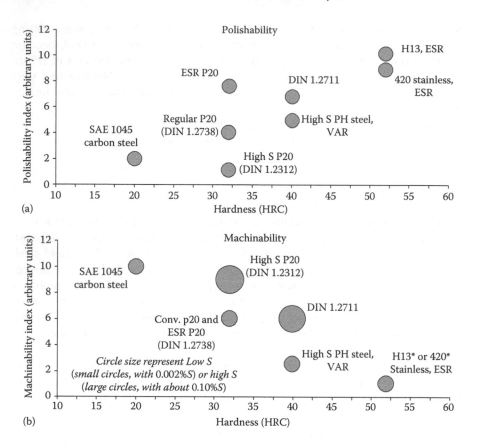

FIGURE 6.10 Effect of hardness on (a) polishability and (b) machinability. *Normally machined before final heat treatment (annealed, 20 HRC), to improve machinability. In (b), the S content is illustrated by the circle size. (Modified from Mesquita, R.A. and Schneider, R., *Exacta*, *São Paulo*, 8(3), 307, 2010.)

The most common texturing process involves preferential etching, which is schematically illustrated in Figure 6.11. In this process, a protection film is first deposited on the mold surface containing the negative of the desired pattern (Figure 6.11a). After deposition on the entire mold surface that comes into contact with the injected part (cavity)—the complexity of which sometimes requires considerable manual labor—the surface + film is placed into an acid tank for etching. During acid etching, the unprotected parts of the cavity surface are corroded, with the removal of material and creation of deeper areas, while the areas under the film are protected from corrosion (Figure 6.11b). When the etching achieves the desired depth, which represents the peaks in the final texture, the etching process is stopped by removing the cavity from the tank and cleaning it. Finally, the film is mechanically or chemically removed, and the cavity is ready for the injection process (Figure 6.11c).

Although the process seems simple, a number of aspects are important to the production of the actual texture. The process as a whole may be then understood as

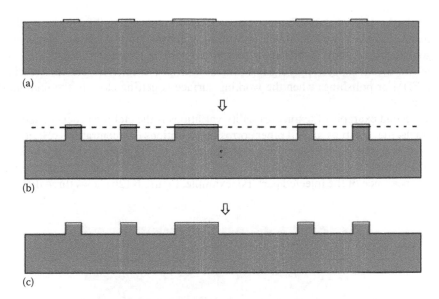

(a)

⇩

(b)

⇩

(c)

FIGURE 6.11 Schematic example of the texturing process applied to a plastic mold surface. After the etching, the film is removed and the mold will contain a topographic pattern exactly the same as the deposited film. The more uniform the corrosion process, the higher the uniformity of the textured surface. (a) Mold surface with deposited film (gray). (b) Mold and film after acid etching process. (c) Final mold surface. (Adapted from Hippenstiel, F. et al., *Handbook of Plastic Mould Steels*, Edelstahlwerke Buderus AG, Wetzlar, Germany, 2001, pp. 17, 40, 67, 125–149, 248–251.)

a "controlled corrosion," which, as in any corrosion process, will depend on (1) the geometrical/surface conditions, (2) the corrosive media, and, finally, (3) the material that is being corroded. It is very important that these three points are always considered, because it is often seen that the tool steel is blamed for poor texturing quality. "This steel is not etching well," it is common to hear, while the correct allegation would be "the result of etching is not good," being related to one or a combination of the three aspects mentioned here. In many cases, a poorly textured surface may not be easily quantified, but it can be only visually estimated on the mold surface. This means that the plastic parts will probably reproduce and have their visual appeal compromised. To explore all possibilities of texturing issues, the three possibilities raised above will be considered next.

The first group of issues in texturing is surface macro irregularities on the mold, prior to the etching process. As in a metallographic preparation, where poor grinding, sanding, or polishing will affect the final etching when observing a microstructure, the plastic mold etching response during texturing will greatly depend on the process applied in the mold finishing. Poorly prepared surfaces or defects on the mold surface may cause distinct corrosion and may visually affect the surface of the textured mold. This will almost certainly affect the produced plastic parts. In several cases, scratches or grooves that were unseen after sanding may even create lines similar to segregation bands after etching. In other cases, like the example in Figure 6.12a, damaged EDM areas may exist, which may turn to be very detrimental to a texturing etching. While many other and specific cases may exist, the common solution is always applying slow and careful processes (conventional machining, EDM or polishing) when the working surface is getting close to the final mold surface.

A second example of texturing quality variations is the etching process itself. For example, the acid type and all other corrosion conditions (concentration, pH, Fe concentration in the tank, temperature, bath homogeneity, etc.) have a substantial effect on the etching process and, consequently, on the final quality of the mold surface and appearance of the injected part. For example, Figure 6.12b shows three different

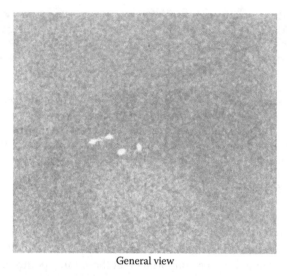

General view

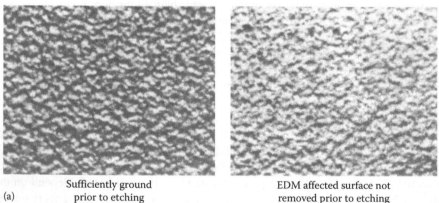

(a)

Sufficiently ground
prior to etching

EDM affected surface not
removed prior to etching

FIGURE 6.12 Defects observed after texturing process. (a) Whites spots after insufficient removal of EDM-affected surface. (From Harberling, E. et al., *Thyss. Edel. Tech. Bere.*, 73, 1990.)   (*Continued*)

patterns applied to the same mold steel under different acid conditions. In addition to the visual appeal, those corrosion differences may also affect the roughness of the valley areas on the textured surfaces, which will later become the top areas on the surface of the injected plastic parts. The reflection of light (shininess) changes with the modification of the corrosive medium, but also the surface and then the touch-sensitive aspect (roughness). To avoid problems, the control of all corrosion parameters, including the fluid flow in the etching tank, is important.

The third group of texturing quality problems refers to the tool steel itself, more precisely the metallurgical features of the tool steel on the mold surface. The common point in terms of mechanism is related to changes in the microstructure, which

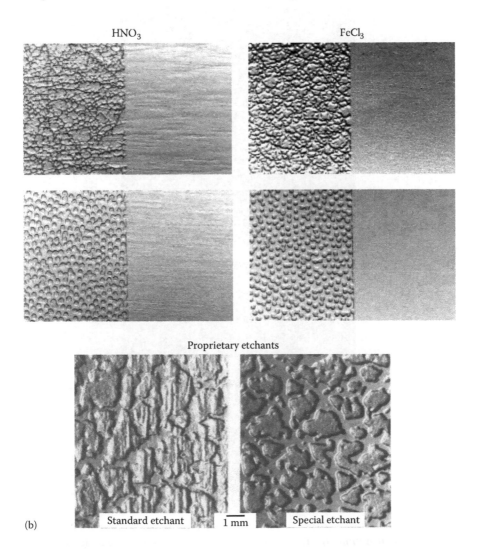

FIGURE 6.12 (*Continued*) Defects observed after texturing process. (b) Etchant effect [1,11]; top image with both grained (left) and polished plus etching (right). (*Continued*)

affects corrosion resistance of the steel and thus modifies the photoetching process. One example is changing the steel grade itself, as displayed in Figure 6.12c, which will immediately affect the heat treatment and the corrosion response during texturing. The corrosive conditions during texturing should be adjusted to each kind of steel and ideally also adapted to (very) distinct heat treatment. Another common change in the microstructure is that produced by thermal processes, mainly by welding (Figure 6.12d). Used to correct small changes or mistakes in mold manufacturing, welding also changes the amount of carbon and alloy elements that are in solid

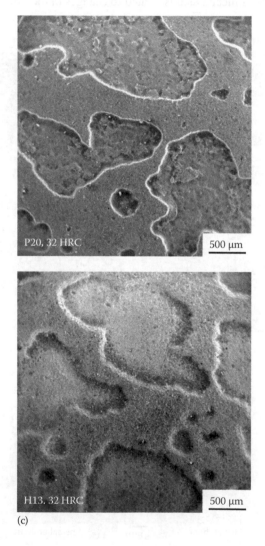

(c)

FIGURE 6.12 (Continued) Defects observed after texturing process. (c) Mold steel chemical composition (From Mesquita, R.A. and Schneider, R., *Exacta*, 8(3), 307, 2010.)

(Continued)

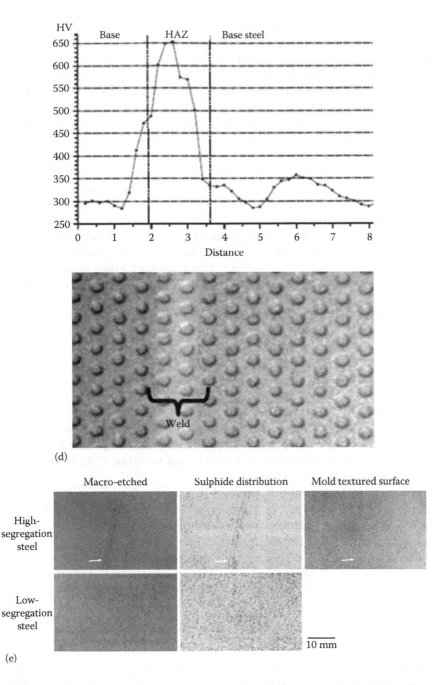

(d)

(e)

FIGURE 6.12 (Continued) Defects observed after texturing process. (d) Welding (From Preciado, T. and Bohorquez, C.E.N., Reparo por soldagem de moldes de injeção de plásticos fabricados em aços AISI P20 e VP50IM, *Proceedings of the 3 Cobef, 2005*, Joinville—SC. April 2005, CDROM.) (e) Macrosegregation in the tool steel, indicated by the arrows. (Adapted from Harberling, E. et al., *Thyssen Edelstahl Technische Berichte*, 16, 73, 1990.)

solution, in addition to changing the microstructure and forming martensite in a small area. Thereby, the corrosion resistance of welded areas is often increased, reducing the etching effect (note the lighter appearance of the welded area in Figure 6.12d). One solution in welded cases is applying a new tempering after welding, which tends to homogenize the microstructure and hardness, thus leading to better homogeneity in the texturing response. Another possibility is applying steels less sensitive to martensite transformation (lower carbon), especially those that may be tempered at high temperature, leading to less hardness (and consequently microstructure) variations.

Another way to promote changes in the microstructure and consequently to the corrosion behavior is by the presence of segregation. As explained in Chapter 2 and in Section 6.2, segregation means the heterogeneous distribution of carbon and alloying elements, those differences being in 0.1 mm range for microsegregation and several millimeters for macrosegregation. One example of a segregation effect on a textured surface is shown in Figure 6.12e, but can also be noted in Figure 6.12b, where the polished and etched surfaces show clear segregation lines from the tool steel blocks. However, it is important to note in Figure 6.12b how the acid conditions also affect the appearance of the segregation lines; in other words, the steel quality is the same for all cases, but the segregation lines are more pronounced or less visible depending on which texture is applied and which acid is used. This fact is important because, especially in tool steels from big molds (above 10 or 20 tons), it is physically impossible to produce conventionally cast ingots without segregation. In some cases, remelted ingots may be used, but the cost usually restricts this route. Solutions to reduce macrosegregation and optimize texturing response have been developed by many steel mills, in terms of new grades, with usually less carbon contents. However, even in those steels where the tendency for segregation is smaller, it is always important to accept that there is a limit to reducing segregation; then the corrosion medium during texturing has to be optimized to avoid highlighting the (natural) segregation on the mold surface.

In summary, the deterministic approach in texturing should be replaced by a more holistic approach, where not only one parameter is considered alone. Rather, the three main groups of texturing variables should be always analyzed together: surface irregularities prior to texturing, such as grinding or sanding defects; the applied acid conditions, type of acid, and corrosion conditions; and finally the tool steel characteristics, including chemical composition, heat treatment, and microstructural heterogeneities caused, for example, by welding or segregation.

6.3.4 Technologies to Improve Plastic Mold Steel Microstructural Cleanliness

From the previous sections, the characteristics of inclusions are seen to be deeply related to the polishing, machining, and photoetching behavior of plastic mold steels. For the purpose of reducing the amount of inclusions, special melting shop processes may be applied in the production of mold steels. This was previously discussed in Chapter 2, and a specific approach is provided here for plastic mold steels. For the most extreme cases of controlling inclusions, the applied processes are related to remelting technologies. This is when an ingot (named electrode in the remelting processes) with the nearly final steel composition is remelted in an environment that enables the reduction of inclusions, followed by solidification with a fast cooling rate, leading to refined microstructures.

The use of remelting processes, especially ESR, in the production of high-quality tool steels found early applications in the 1970s [13] and has been constantly improved since then [14–19]. While the focus during the first decades was on low sulfur contents (and corresponding improvements in toughness properties) and covered high-speed steels, cold and hot work tool steels alike [13,20], the focus has recently turned to hot work tool steels and plastic mold steels [21–24].

Besides the standard (open) ESR processes, pressure- and/or protective-gas ESR (P-ESR) has found widespread application in the last decade as an improved process alternative [15–19,25]. Furthermore, VAR—usually applied to steels and Ni-based superalloys in the aircraft industry—has also gained some share in the production of tool steels [18,21].

Table 6.2 gives an overview on the advantages of and the differences between the different remelting processes. Besides these material-related effects, there are also some differences in the handling of the ingots. For instance, VAR ingots require some surface conditioning, and open ESR permits a more flexible ingot weight [16].

The main advantages of P-ESR in comparison to standard ESR are the better material properties due to reduced segregations and nonmetallic inclusions, and the possibility of producing high-nitrogen alloyed steels. VAR materials have the highest homogeneity regarding segregations and cleanliness, but have the risk of white spots (macro defects from the remelting process). Furthermore, the lowest contents

TABLE 6.2

Advantages of and Differences between the Different Remelting Processes

	ESR	P-ESR	VAR
High ingot homogeneity	Yes	Yes	Yes
Microsegregations	Low	Lower	Lowest
White spots	No	No	Yes
Cleanliness (nonmetallic inclusions)	Good	Better	Best
N alloying	No	Yes	No
N-content reduction	Limited	No	Yes
H content	Increasing	Constant	Decreasing
Low Si and Al contents	No	Yes	Yes
Reduction of trace elements	No	No	Yes

Source: Mesquita, R.A. and Schneider, R., *Exacta*, 8(3), 307, 2010.

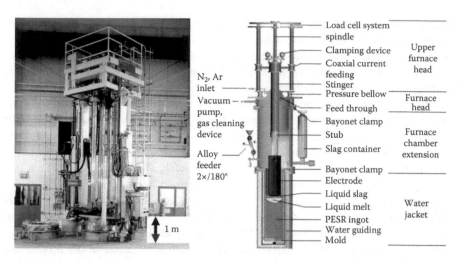

FIGURE 6.13 Appearance and configuration of modern P-ESR plant. (From Koch, F. et al., *Proceedings of the Fourth Symposium on Advanced Technologies and Processes for Metals and Alloys*, Frankfurt, Germany: DGM. June 1999, pp. 47–50.)

in volatile trace elements such as antimony, arsenic, or tin, as well as H and N contents, can only be achieved by this vacuum process. The typical appearance and configuration of a modern P-ESR plant according to Koch et al. [16] and other authors [17–19,26,27] can be seen in Figure 6.13.

The main advantages of remelted materials for application in the tooling industry are their superior homogeneity and isotropy of mechanical properties, especially regarding toughness, and the outstanding polishability for the most demanding surface requirements [13,20]. Toughness is mainly affected by segregations, which range from trace elements to carbide stringers at grain boundaries. Besides the improvement of the solidification conditions regarding direction and local solidification time, which are dominant in hot work tool steels, the alloying of nitrogen has proven to be very beneficial for improving the microstructure of corrosion-resistant plastic mold steels [13,16,17,20].

Severity level number	Inclusion type							
	A		B		C		D	
	Thin	Thick	Thin	Thick	Thin	Thick	Thin	Thick
0.5	(a)	(a)	(b)					
1.0	(a)		(b)					
1.5							(b)	
2.0								

FIGURE 6.14 Limits regarding nonmetallic inclusions for high-quality tool steels; hatching, possible occurrence; dark gray, not observed with proper remelting. (a) Depending on prior steel desulfurization; (b) Depending on ladle metallurgy. (From Schneider, R. et al., *BHM Berg Hüttenmänn. Monatsh.*, 145(5), 199.)

While the sulfur content can be reduced to very small levels by modern ladle metallurgy, the lowest contents in oxides can be achieved only through remelting. Thus, good ladle metallurgy is again an important precondition for satisfactory results after remelting. Higher levels of cleanliness can be achieved through by ESR under a protective atmosphere in combination with adapted slag compositions and through by VAR [16,21,25]. The typical limits for high-quality tool steels regarding nonmetallic inclusions, according to ASTM E45 - Method A, can be taken from Figure 6.14. Thereby, areas indicated by "a" and "b" will be affected by lathe metallurgy processes.

6.4 MAIN CHARACTERISTICS AND SELECTION OF PLASTIC MOLD STEELS

From the previous discussion, it is clear that the hardness and the resultant wear resistance are important properties of the mold steels, but most properties are actually a response from the steel to the manufacturing processes applied. In this respect, a chart of comparative properties of mold steels was prepared, and a discussion is included in the following on the main plastic steels, divided into five groups: (1) the modified P20 grades, supplied in the prehardened condition with 32 HRC; (2) 40 HRC prehardened grades, such as DIN 1.2711; (3) H13 steel, heat-treated to hardness above 50 HRC to improve polishability; (4) corrosion-resistant grades; and (5) high-hardness/high-wear-resistance steels, based on cold work or powder metallurgy (PM) grades (Figure 6.15).

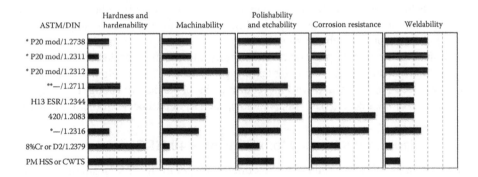

FIGURE 6.15 Main properties and manufacturing characteristics of plastic mold steels. * Prehardened to 32 HRC; **Prehardened to 40 HRC. For the other steels, the machining considers the supply in the annealed condition and the other characteristics are for final heat treatment to the most common hardness in plastic mold tooling: H3 and 420 to 52 HRC, cold work steels to 60 HRC, and PM to 65 HRC. (Comparative diagrams were drawn with information from Hippenstiel, F. et al., *Handbook of Plastic Mould Steels*, Edelstahlwerke Buderus AG, Wetzlar, Germany, 2001, pp. 17, 40, 67, 125–149, 248–251; Uddeholm Tooling, Technical catalogue for tool steels, Available online at: http://www.uddeholm.com/files/, Accessed December 2015; Böhler Edelstahl, Technical catalogue for tool steels, Available online at: http://www.bohler-edelstahl.com/english/1853_ENG_HTML.php, Accessed December 2015; Villares Metals, Technical catalogue for tool steels, Available online at: http://www.villaresmetals.com.br/pt/Produtos/Acos-Ferramenta, Accessed December 2015.)

6.4.1 Modified P20 Grades

In terms of the volume and the number of molds used in the plastic industry, the modified P20 grades are by far the most important group. Three main steels may be included in the category of modified P20, all from the DIN standards and well known by their DIN numbers: 1.2738, 1.2311, and 1.2312. The first grade, 2738, is the most important among the three and so the main representative of plastic mold steels. It has a good balance between hardness and, through hardening (hardenability), good machinability, and good polishability/etchability. The cost, especially in large blocks, is an important factor and another advantage of 2738 in relation to higher alloyed grades. The microstructure, as explained earlier, will be fully bainitic after hardening with oil quenching and tempering, leading to a smaller decrease in hardness in the core areas of large blocks. As previously mentioned, large blocks have to be treated with caution in the finishing operation, such as adjusting the etching during texturing, because they will always contain a certain amount of macrosegregation.

When the penetration of hardness is not a strong requirement, such as in blocks with smaller thickness or where when cavities are not deep, 2311 grade may be applied in substitution of 2738. The main difference between those grades is the maximum depth of hardening, after quenching, in a way that large blocks may have a low core hardness if made of 2311 (Figure 6.5). But for smaller blocks (thickness less than 400 mm), 2738 and 2311 behave similarly. And the low Ni amount in 2311 brings some cost reduction in relation to 2738.

As for the grade 2312, the same rationale is valid in terms of size limitation, with the main differences including machinability and polishability or ethability. Grade 2312 has high contents of S (~0.1%), which increases the machinability and decrease the machining times for mold rough machining due to the presence of large amounts of MnS inclusions, which are soft and help the machining process. However, these inclusions also cause pin-point defects after polishing and may accelerate the over-polishing and also affect the etching for texturing. Therefore, 2312 is often used in molds without deep engraving and with no requirements of surface finishing, typically for back areas of the injected parts (male molds), such as in bumpers or other large parts.

6.4.2 Prehardened Plastic Mold Steels with 40 HRC

Steels such as 1.2711 and others in the same family, or even 2738 for smaller dimensions, may be hardened and tempered to a final hardness of 40 HRC and supplied as blocks with this condition. The machining technologies today enable the manufacturing under this condition by controlling the cutting parameters. And the higher hardness, compared to the 32 HRC of modified P20, leads to advantages in terms of mold design and the polished quality. The last aspect of shinier surfaces is important for several automotive and household appliances applications.

Long drilling of the cooling channels may be an issue with this hardness, especially when macrosegregation aspects are involved and harder areas may exist. Welding is also an issue in most cases, due to the high carbon and crack sensitivity after welding. Therefore, some steel suppliers offer alternatives with 40 HRC but lower carbon contents to solve both the segregation and the welding issues.

Usually those options come with a price increase, which in many cases is not back-calculated in relation to the cost reduction in the processing steps of machining, drilling, or welding. The modified plastic mold steels with 40 HRC are not the majority, and still grades such as 1.2711 and 1.2714 are very popular in this market.

6.4.3 ESR-Refined H13 Tool Steel for Plastic Mold Applications

For very highly polished parts, such as automotive lenses and highly polished molds, high alloy steels may be used. The high hardness after hardening and tempering and the availability ESR-refined blocks make H13 very popular in those applications. In some cases, hard coatings or nitriding is added to the mold to improve the wear resistance or polishing characteristics. The high tempering resistance (see Chapter 4) of H13 steel makes it suitable those surface treatments, without the risk of a loss in core hardness. The material cost, however, limits the application of ESR H13 to very specific molds, where the polishing requirements are very high. The main cost changes are due to the high material cost, in comparison to P20, and the more complex (and expensive) heat treatment, usually performed after rough machining and in vacuum furnaces.

6.4.4 Corrosion-Resistant Plastic Molds

Some plastic mold steels may require high corrosion-resistant grades due to the storage in areas exposed to corrosion, but mainly to the processing of chlorinated polymers such as polyvinyl chloride (PVC). During this thermoplastic processing, the Cl in the polymer structure may be released by degradation processes and combine with vapor, forming HCl, which will attack and corrode the metallic parts, including the molds. In other cases with very high polishing requirements, corrosion resistance may be required only as a guarantee that there will be no damage to the mold surface when exposed, due to corrosion, because high manufacturing cost may have been added during the finishing of the mold.

In those situations, stainless steels are used, usually high-Cr steels, typically with 12%Cr or more. ASTM 420 mod or DIN 1.2083 is the most common, but other grades are also found in the market, such as 1.2316 or modifications thereof. Especially for 2316, the supply conditions may be specified as prehardened to 32 HRC, which reduces the mold manufacturing complexity and leads to faster production by not involving a heat treatment between rough machining and finish machining. However, to increase polishability and also corrosion resistance, stainless steels should be heat-treated to a higher hardness, above 50 HRC. And, for martensitic stainless steels such as grade 420, this hardness level may be achieved either by tempering at a low temperature (250°C) or a high temperature (~500°C). The higher tempering conditions involve strong precipitation of Cr secondary carbides, depleting the Cr in solid solution and reducing corrosion resistance. Therefore, steel manufacturers usually recommend using a low temperature for tempering martensitic stainless steels, such as 420 or similar grades (see Reference 28). Proprietary composition also exists for

2

corrosion-resistant molds, some based on low carbon precipitation hardening steels. While those show better weldability and corrosion resistance, their cost and lower machinability are drawbacks.

6.4.5 WEAR-RESISTANT STEELS FOR PLASTIC APPLICATIONS

Although rare, there are plastic injection processes that may cause wear on the mold steel. This happens because ceramic or fiber materials are often added to the plastic mixture in order to increase mechanical properties of the plastic part, such as strength or stiffness. However, during the flow into the mold, those fibers or particles may lead to abrasion on the mold surface. To prevent this, the solutions are similar to what was discussed in hot work and cold work steels: increase the matrix hardness and/or use tool steels with undissolved carbides in their microstructure.

With this approach, dozens of different steels may be used in abrasive plastic processing, including high-hardness steels such as H13 or nitrided H13, typical cold work tool steels like D2 or 8%Cr steels, and standard or new PM steels. The last option, although involves higher cost, is interesting in terms of properties, because the small carbides do not lead to a strong decrease in polishability. In addition, some stainless steel abrasion-resistant PM grades also exist in the market and may be an option for combined corrosion- and abrasion-resistant plastic injection processes.

REFERENCES

1. F. Hippenstiel, V. Lubich, P. Vetter, W. Grimm. *Handbook of Plastic Mould Steels.* Wetzlar, Germany: Edelstahlwerke Buderus AG, 2001, pp. 17, 40, 67, 125–149, 248–251.
2. ÖNORM EN ISO 4957:1999. *Tool Steels.* Publisher and printing: Österreichisches Normungsinstitut, 1020 Wien.
3. ASTM A681-08. Standard specification for tool steels alloy. Materials Park, OH: ASTM International.
4. Uddeholm Tooling. Technical catalogue for tool steels. Available online at: http://www.uddeholm.com/files/. Accessed December 2015.
5. Böhler Edelstahl. Technical catalogue for tool steels. Available online at: http://www.bohler-edelstahl.com/english/1853_ENG_HTML.php. Accessed December 2015.
6. Villares Metals. Technical catalogue for tool steels. Available online at: http://www.villaresmetals.com.br/pt/Produtos/Acos-Ferramenta. Accessed December 2015.
7. M.C. Flemings, G.E. Nereo. Macrosegregation: Part 1. *Transactions of the Metallurgical Society of AIME*, 239, 1449–1461, 1967.
8. T. Shimizu, T. Fuji. Mirror surface finishing properties of plastics mold steels. *Electric Furnace Steel, Japan*, 74(2), 125–130, 2003 (Daido Steel Co. Ltd., Osaka, Japan).
9. Uddeholm Technical Datasheet. Polishing mould steel. Available at: http://www.uddeholm.com/files/polishing-english.pdf. Accessed in March 2016.
10. R.A. Mesquita, R. Schneider. Tool steel quality and surface finishing of plastic molds. *Exacta, São Paulo*, 8(3), 307–318, 2010.
11. E. Harberling, G. Mazza, A. Moriggi, K. Rasche, B. Remi. Faults during photo-etching of plastic mould steels. *Thyssen Edelstahl Technische Berichte*, 16, 1990, 73–81.
12. W.T. Preciado, C.E.N. Bohorquez. Reparo por soldagem de moldes de injeção de plásticos fabricados em aços AISI P20 e VP50IM. *Proceedings of the 3 Cobef, 2005*, Joinville—SC, Brazil. April 2005, CDROM.

13. E. Krainer, W. Holzgruber, R. Plessing. Praktisch isotrope Werkzeugst.hle und Schmiedestücke h.chster Güte. *BHM Berg- und Hüttenm.nnische Monatshefte*, 116(3), 78–83, 1971.

14. R. Hasenhündl et al. Herstellung, Eigenschaften und Anwendungen ausgewählter korrosionsbeständiger, stickstofflegierter Stähle. *Proceedings of "Forum für Metallurgie und Werkstofftechnik,"* Leoben, Austria: Austrian Metallurgy Society, vol. 11–12, May 2009, pp. 35–36.

15. C. Heischeid, H. Holzgruber, and A. Scheriau. Advanced production technology for large ESR ingots. *Proceedings of the Seventh International Tooling Conference—Tooling Materials and Their Applications from Research to Market*, Torino, Italy, Eds. M. Rosso Actis Grande, D. Ugues, May 2–5, 2006, vol. 2, 2006, pp. 423–430.

16. F. Koch, P. Würzinger, R. Scheneider. Advanced equipment for the economic Production of speciality steels and alloys. *Proceedings of the Fourth Symposium on Advanced Technologies and Processes for Metals and Alloys, 1999*, Frankfurt, Germany: DGM Frankfurt, June 1999, pp. 47–50.

17. R. Schneider, F. Koch, P. Würzinger. Pressure-electro-slag-remelting (PESR) for the production of nitrogen alloyed tool steels. *Proceedings of the Fifth International Tooling Conference "Tool steels in the next century"*, Eds. F. Jeglitsch, R. Ebner, H. Leitner, Leoben, Austria, September 29–October 1, 1999, pp. 265–273.

18. R. Schneider, P. Würzinger, G. Lichtenegger, H. Schweiger. Metallurgie an den technischen Grenzen höchster Reinheitsgrade und niedrigster Spurenelementgehalte. *BHM Berg- und Hüttenmännische Monatshefte*, 145(5), 199–203, 2000.

19. R. Schneider, F. Koch, P. Würzinger. Metallurgical advances in pressure ESR (PESR). *Proceedings of the International Symposium on Liquid Metal Processing and Casting*, Santa Fe, NM, September 23–26, 2001, pp. 105–117.

20. H.J. Becker, E. Haberling, K. Rasche. Herstellung von Werkzeugstählen durch Elektro-Schlacke-Umschmelz- (ESU-) Verfahren. *Thyssen Edelstahl Technische Bereichte*, 15(2), 138–146, 1989.

21. H. Schweiger, H. Lenger, H.-P. Fauland, K. Fisher. A New generation of toughest hot-work tool Steels for highest requirements. *Proceedings of the Fifth International Tooling Conference "Tool Steels in the Next Century"*, Leoben, Austria, Eds. F. Jeglitsch, R. Ebner, H. Leitner, September 29–October 1, 1999, pp. 285–294.

22. G. Lichtenegger, R. Schneider, J. Sammer, G. Schirninger, P. Würzinger, J. Neuherz. Development of a nitrogen alloyed tool steel. *Proceedings of the Fifth International Tooling Conference "Tool Steels in the Next Century"*, Leoben, Austria, Eds. F. Jeglitsch, R. Ebner, H. Leitner, September 29–October 1, 1999, pp. 643–652.

23. K. Sammt, J. Sammer, J. Geckle, W. Liebfahrt. Development trends of corrosion resistant plastic mould steels. *Proceedings of the Sixth International Tooling Conference "The Use of Tool Steels: Experience and Research"*, Karlstad, Sweden, Eds. J. Bergström, G. Fredriksson, M. Johansson, O. Kolik, F. Thuvander, September 10–13, 2002, pp. 285–292.

24. R. Schneider, R. Pierer, K. Sammt, W. Schützenhöfer. Heat treatment and behaviour of new corrosion resistant plastic mold steel with regard to dimensional change. *Proceedings of the Fourth International Conference on Quenching and the Control of Distortion*, Beijing, People's Republics of China, November 23–25, 2003, pp. 357–362.

25. R. Hasenhündl, S. Eglsäer, H. Orthaber, R. Tanzer, W. Schützenhöfer, R. Schneider. Herstellung, Eigenschaften und Anwendungen ausgewählter korrosionsbeständiger, stickstofflegierter Stähle. *Proceedings (Tagungsunterlagen) of the "Forum für Metallurgie und Werkstofftechnik"*, Leoben, Austria, May 11–12, 2009, pp. 35–36.

26. R. Schneider, K. Sammt, R. Rabitsch, M. Haspl. Heat treatment and properties of nitrogen alloyed martensitic corrosion resistant steels. *Transactions of Materials and Heat Treatment—Proceedings of the 14th IFHTSE Congress*, Shanghai, People Republic of China, vol. 25(5), October 26–28, 2004, pp. 582–587.
27. J. Korp, R. Schneider, R. Hasenhündl, P. Presoly. Electrical conductivity of new MgO containing ESR slags and its effect on energy consumption. *Proceedings of the 16th IAS Steelmaking Conference*, Rosario, Argentina, November 5–8, 2007, pp. 333–342.
28. R.S.E. Schneider, R.A. Mesquita. Advances in tool steels and their heat treatment, Part 2: Hot work tool steels and plastic mould steels. *International Heat Treatment and Surface Engineering*, 5, 94–100, 2011.

7 High-Speed Steels

High-speed steels are perhaps the most complex class of tool steels, combining the characteristics of hot work and cold work tool steels. While their requirements of wear resistance and high hardness are similar to those of cold work steels, high-speed steels also have extremely high hot strength and tempering resistance, and in this aspect they are close to hot work steels. Therefore, we recommend reading Chapters 4 and 5, dedicated to cold and hot work steels, respectively, before this chapter. Nevertheless, several references are made to the previous chapters, to remind the reader of the important aspects that have been treated previously. In addition, Chapter 3 will also be referred to extensively during the discussions on high-speed steels because the complex metallurgical reactions of this class of steels are deeply based on most of the physical metallurgy and heat treatment phenomena detailed in Chapter 3.

In spite of the complexity, this chapter is shorter than the previous ones. First, because a large part of the content is already presented in Chapters 3 through 5, these are only referred to here, especially when considering the concepts and data on undissolved carbides, precipitation hardening, and retained austenite. Second, because of the decrease in application of high-speed steels. Being the main application filed in cutting tools for machining processes, the use of high-speed steel tends to decrease because of extensive substitution by other cutting tool materials such as cemented carbides. High-speed steels may also be employed in molds and dies, but the possibilities are limited by cost considerations, specifically material and heat treatment costs. The main applications of high-speed steels in molds and dies were already discussed in Chapters 4 and 5.

Therefore, the focus of the present chapter is, briefly, to show the aspects of physical metallurgy and applications of high-speed steels, mainly in cutting tools. A general review on the requirements for cutting tools is first presented, followed by heat treatment and its effect on their high-temperature strength. Finally, the carbide distribution aspects in high-speed steels are considered.

7.1 INTRODUCTION TO CUTTING TOOL MATERIALS

To understand the application of high-speed steels in cutting tools, other materials and the main properties required for this application also have to be considered. Starting from the first aspect, the different materials for cutting tools, including nonmetallic materials, are presented in Figure 7.1a, where the main properties are addressed: the transverse rupture strength, the fracture toughness (K_{IC}), and the hardness. While hardness is directly related to the wear resistance, the other two properties (rupture strength and K_{IC}) are connected to the toughness or cracking resistance of the tool material. In simpler terms, Figure 7.1a indicates the relationship that exists between

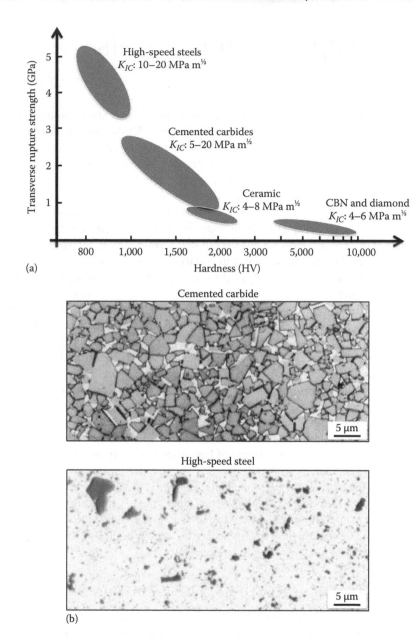

(a)

Cemented carbide

High-speed steel

(b)

FIGURE 7.1   General comparison of materials used for cutting tools. (a) Diagram showing the bend strength, hardness, and K_{IC} values for the different classes of materials. (b) Examples of the microstructures for cemented carbides and high-speed steels, showing the larger amount of carbides (in gray) for the cemented carbides. (Chart built with data from Trent, E.M. and Wright, P.K., *Metal Cutting*, 4th edn., Butterworth-Heinemann, Boston, MA, 2000, pp. 132–249.) The image of cemented carbide is adapted from Reference 1. High-speed steel is from unpublished material from the author, where the carbides were viewed by deep etching and the image was color-inverted, in such a way that the white carbides become gray and the matrix white.

wear resistance and the drawback in toughness* for all cutting tool materials in use today, from high-speed steel to diamond.

Although in extreme conditions ceramic tools, harder tools such as CBN (cubic boron nitrides), or even industrial diamond may be used, the most popular cutting tool materials are cemented carbides, especially those based on tungsten carbide (WC). The microstructure of those cemented carbides is then constituted by a small amount of metallic phase, such as Co for WC, which is used to bind the carbides during tool manufacture (by high-temperature pressing) and to increase the toughness of the cemented carbide. The final amount of metallic phase is small, usually between 5% and 20% in volume. For high-speed steels, however, the situation is the opposite, where only 5%–20% volume of the microstructure is constituted by carbides, while the rest is a hardened metallic matrix. This difference is clear in Figure 7.1b, where the microstructures for both materials are shown.†

Hardness and wear resistance alone, however, does not determine the cutting performance of a tool, because the effects of heat release during machining and the "thermal damage to the tools" have to be considered. In fact, before the emergence of high-speed steels in the early 1900s, the main tools used were composed of high-carbon low-alloy steels, which were hardened to high hardness at room temperature but, not having enough tempering resistance, could not resist the heating process during machining. Alloying with W made a new class of steels able to machine at higher speeds; so, those materials were called *high-speed steels*. After World War II and the shortage of W, Mo was tested as a substitute for W, showing similar characteristics of tempering resistance and the formation of carbides, and then two different (sub)classes of tool steels were created: the tungsten type (T) or the molybdenum type (M) of high-speed steels. While both still exist, the M2 type is today the most common in industrial use, with a balance of W and Mo contents. For reference, Table 7.1 presents the chemical composition of the most common high-speed steels.

Moving back to the heat effect in cutting tools, Figure 7.2 shows several examples on how the machining parameters affect the generation of heat and increase the tool's temperature, which in turn leads to acceleration of the cutting-edge wear (Figure 7.2a). In fact, for different tool geometries, the main machining parameter, that is, the cutting speed, directly affects the heating of the cutting tools, as can be seen in Figure 7.3b. This in turn brings the need for materials that are not only hard but also show a minimum stability of hardness with the increase of temperature; Figure 7.2c shows this for three groups of cutting tool materials: carbon steels, high-speed steels, and cemented carbides.

As mentioned in the previous paragraph, this increase in temperature resistance and the consequent achievement of higher speed in machining was essential for the development and application of high-speed steels. Section 7.2 will consider this

---

* This inverse relationship between hardness and toughness follows the natural inverse relation between both properties, which is observed in many classes of metallic and nonmetallic materials. The reason is that the increase in hardness usually is accompanied by a decrease in the ability to deform plastically (or ductility), which is important to absorb energy. As a consequence, the harder the material, the lower the observed toughness.
† A special etching and image treatment was applied to enable the high-speed steel also to show the undissolved carbides as gray particles.

TABLE 7.1

Chemical Composition of the Main High-Speed Steels, Showing the Commonly Observed Target Values

High-Speed Steels			Most Common Chemical Composition (wt.%)							
ASTM	EN/DIN	UNS	C	Si	Mn	Cr	V	W	Mo	Other
M1	1.3346	T11301	0.82	0.3	0.3	4.0	1.0	1.5	8.0	—
M2	1.3343	T11302	0.89	0.3	0.3	4.0	1.9	6.0	5.0	—
M3:2	1.3344	T11323	1.20	0.3	0.3	4.0	3.0	6.0	5.0	—
M35	1.3243	T11335	0.89	0.3	0.3	4.0	1.0	6.0	5.0	Co = 5.0
M42	1.3247	T11342	1.10	0.3	0.3	4.0	1.1	1.5	9.5	Co = 8.0
M50	1.3551	T11350	0.84	0.4	0.4	4.0	1.0	—	4.2	
T1	1.3355	T12001	0.75	0.3	0.3	4.0	1.0	18	—	—
T15	1.3202	T12015	1.55	0.3	0.3	4.0	5.0	12	—	Co = 5.0

Note: The specific limits and ranges are given by the standards ASTM A600 and Euro Norm ISO 4957 (old DIN 17350), according to References 2 and 3. M3:2 refers to the class 2 of M3 in ASTM standard.

high-temperature (or tempering) resistance of high-speed steels, as was explained in Section 3.2.2.3. On the other hand, the same effect of higher temperature hardness is also the reason why other materials such as cemented carbides or ceramic tools have been substituting high-speed steels. By being more resistant to tool heating during the cutting process, cemented carbides or other materials enable the application of higher machining parameters, which then improves machining productivity. The applications for which this cannot be done are tools sensitive to cracking or chipping, where high-speed steels are still used extensively (e.g., taps and large gear-machining tools).

7.2   HEAT-TREATMENT RESPONSE IN HIGH-SPEED STEEL

The heat-treatment response of high speed-steels has similar concepts as for hot work tool steels, promoting strong secondary hardening (increase in hardness after tempering above 500°C/930°F) by alloy carbide precipitation. Because of the higher amount of V, Mo, and W, combined with the high carbon content, precipitation hardening in high-speed steels leads to extremely high levels of hardness, in some cases achieving more than 68 HRC. Figure 7.3 shows the tempering diagrams for most high-speed steels, over the complete range of tempering (Figure 7.3a) and, more importantly, within the range used for industrial tempering conditions (Figure 7.3b).

Some observations are important to be made in those diagrams. First, from Figure 7.3a it is clear that the temper hardness increases on increasing the hardening temperature. As explained in detail in Section 3.2.2.1, higher hardening temperatures lead to more dissolution of carbides, which then precipitate with higher intensity during tempering, promoting stronger secondary hardening. A second observation concerns the difference in chemical composition and the obtained temper hardness.

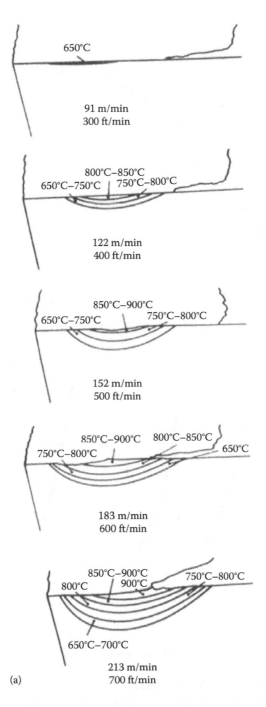

FIGURE 7.2 Effect of cutting parameters on the cutting tool damage. (a) Images of increased tool wear and temperature at the cutting edge when increasing the cutting speed.

(*Continued*)

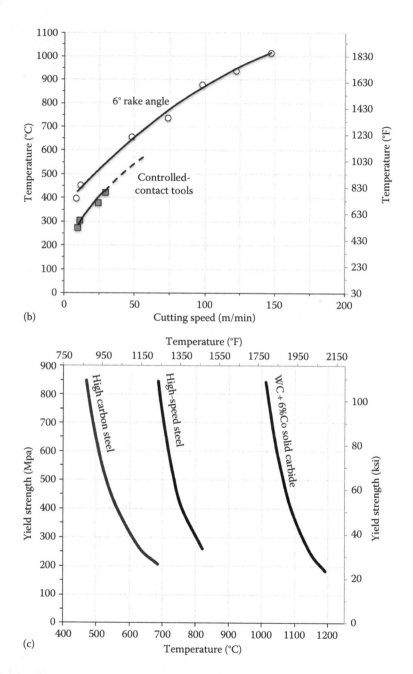

FIGURE 7.2 (*Continued*)   Effect of cutting parameters on the cutting tool damage. (b) Effect of cutting speed on the tool edge maximum temperature. (c) Reduction of strength with the increase in temperature for distinct materials. (All charts modified from Trent, E.M. and Wright, P.K., *Metal Cutting*, 4th edn., Butterworth-Heinemann, Boston, MA, 2000, pp. 132–249.)

Comparing Figure 7.3b to Table 7.1, it is clear that there exist four main groups of high-speed steels:

1. The so-called semi-high-speed steels alloy steels, such as M50, which achieve maximum hardness of 63 HRC because of the leaner chemical composition. This low hardness level is also obtained by hardening traditional grades, such as M2, at lower temperatures, which brings fewer elements into solid solution and ends up being in a similar condition as in a low-alloy high-speed steel.
2. Standard high-speed steels, such as M2 and T1, which achieve temper hardness in the range of 64–65 HRC, as a result the amount of alloying elements, especially Mo, W, and V.
3. High-speed steel compositions achieving extremely high hardness levels, above 65 HRC, often called *super-high-speed steels*. Typical examples are M42 and T15 tool steels. The reason for their extremely high hardness is the presence of a high Co content, which although does not form secondary carbides but increases the effect of precipitation hardening by carbides of other alloying elements (Mo, W, and V).
4. Powder metallurgy (PM) grades, in which the presence of refined carbides (see Section 7.3) leads to higher dissolution during hardening and then stronger precipitation during tempering. There exist several types of PM grades, with proprietary or nonstandardized composition, some with Co to increase the temper hardening. Since the microstructure is refined by the

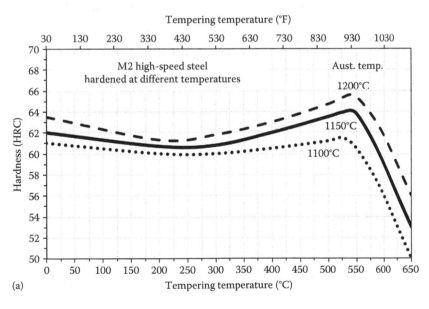

(a)

FIGURE 7.3 Tempering diagrams for the most common cold work tool steels, showing hardness after two times tempering for 2 h. (a) Complete curve, showing the secondary hardening effect for M2 tempered in the full range, with three different hardening temperatures.

(Continued)

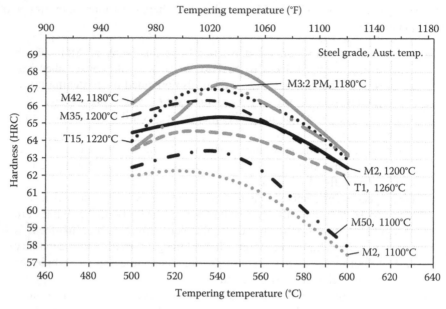

FIGURE 7.3 (*Continued*)   Tempering diagrams for the most common high-speed steels, showing hardness after two times tempering for 2 h. (b) Data from brochures of traditional tool steel producers. (From ASTM A600-92, Standard specification for tool steel high speed, ASTM International, Materials Park, OH; Erasteel, Grade datasheets for high-speed steel products, Available online at: http://www.erasteel.com/content/asp-steels-0, Accessed April 2016; Böhler Edelstahl, Technical catalogue for tool steels, Available online at: http://www.bohler-edelstahl.com/english/1853_ENG_HTML.php, Accessed April 2015.)

fast solidification rate, the hot rolling of PM grades is relatively easy, and so compositions that cannot be produced through conventional casting route are possible by PM. So, the alloying possibilities are much wider in PM grades, leading to stronger secondary hardening by a combination of fine carbides and richer compositions.

A few important characteristics are also unique in the heat treatment of high-speed steels. First, the hardening conditions are very high and close to the solidus temperature (where liquid starts to be formed again). This is due to the achievement of higher dissolution (once solubility is determined by the hardening temperature), but it can generate excessive austenite grain growth. Very short hardening times are then applied in high-speed steels to avoid this excessive grain growth, typically in the range of 2–5 min (at hardening temperature) only. Another recommendation is the use of the minimum necessary hardening temperature, as shown in Figure 7.4a.

In this effect of the hardening temperature, a second aspect of high-speed steel's heat treatment is identified. While most tool steels have a fixed temperature and hardness is adjusted by the tempering conditions, it is recommended that high-speed steels are always tempered slightly above the secondary hardening peak, ~550°C–560°C (~1020°F–1040°F), and the hardness is controlled by changing the amount of dissolution during hardening, meaning that the hardening temperature then determines the temper hardness (Figure 7.4b). For a fixed tempering close to the secondary hardening peak, an increase in the hardening temperature leads to an increase in temper hardness, until a saturation point is reached, mainly due to retained austenite effects.

Toughness decreases continuously with increase in hardness, which may be observed in Figure 7.4b in terms of the rupture strength and energy absorption. This first means that the hardness achieved should be just the required for the application, leading to the maximum possible toughness. In addition, it also means that extremely high hardening temperatures are not effective, since little effect is seen in hardness and consequent wear resistance, but toughness is strongly affected, negatively, when the hardening temperatures are too high. The reason for the decrease in toughness is the natural response due to the increase in hardness, but it also is affected by the grain growth. In other words, since the temper

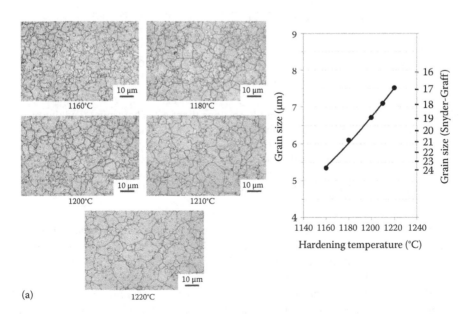

(a)

FIGURE 7.4 Bend test results showing the effect of hardening condition on the microstructure and properties of M2 high-speed steel. (a) Grain size. *(Continued)*

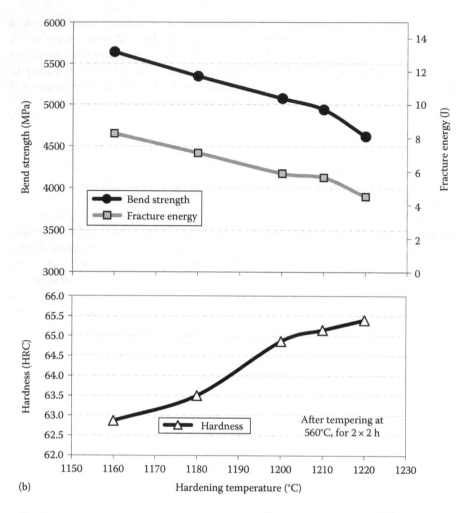

FIGURE 7.4 (Continued) Bend test results showing the effect of hardening condition on the microstructure and properties of M2 high-speed steel. (b) Hardness and bend test results—bend strength and energy absorption (toughness). The increase of hardness is achieved by the most common and recommended practice for high-speed steels: fixing the tempering at around 560°C and increasing the hardening temperature to obtain higher hardness. (Data from Mesquita, R.A. et al., *Int. Heat Treat. Surf. Eng.*, 5, 36, 2011.)

hardness is controlled by hardening, any attempt to obtain higher hardness may lead to little or no effect in the hardness itself, but strong loss in toughness. This is actually observable in Figure 7.4b for the M2 high-speed steel in the case of hardening above 1210°C or 1220°C (2230°F).

Still on this point of controlling the hardness by the hardening temperature, and not by tempering adjustment, another situation may be considered: the application of a high-speed steel in a die with specified hardness of 60–62 HRC. In this case, Figure 7.5a displays two possibilities: hardening at high temperature and tempering

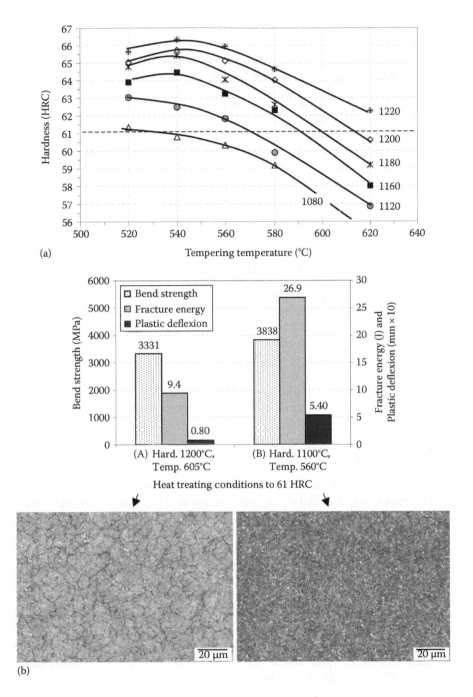

FIGURE 7.5 Example of the two strategies to reduce the hardness in high-speed steels. (a) Keeping high hardening temperature and adjusting the tempering (left). (b) Keeping the tempering a few degrees beyond and adjusting the hardening to lower temperatures. (From Mesquita, R.A. et al., *Int. Heat Treat. Surf. Eng.*, 5, 36, 2011.)

also in high temperature, or hardening in lower temperature and keeping the tempering at around 550°C (1020°F). The results for both will be a hardness of 61 HRC, but as shown in Figure 7.5b, the effect on toughness is very different. Because of a microstructure less exposed to high temperatures during hardening (small prior austenite grain size), the toughness value shows a clear advantage for the second process: reducing the hardening temperature to reduce the hardness, but keeping tempering close to the peak at around 550°C (1020°F).*

In summary, hardening is one of the most important steps in the heat treatment of high-speed steels, being also responsible for the achieved final hardness if tempering is fixed close to the peak of 550°C (1020°F). If very high hardness levels are necessary, there is no other option but to raise the hardening temperatures to the maximum suggested values (given by the manufacturers, but for M2 this is ~1200°C/2190°F). On the other hand, if a lower hardness is necessary, hardening at lower temperatures brings the advantage in toughness by better controlling the grain growth during hardening.

7.3 CARBIDES IN HIGH-SPEED STEELS

Similar to cold work steels, the microstructure that determines the properties of high-speed steels is characterized by two features: a high hardness matrix, tempered at high temperatures and so resistant to heating; and a dispersion of undissolved carbides. The second part is focused on in this section, which basically discusses the types of carbides, with their intrinsic hardness, and the distribution of those carbides in the microstructure. The types of carbides in tool steels and their formation and distribution were covered in detail in Sections 3.3 and 5.3.2. In this section, just a few comments are given on those carbides with regard to the specific points on the high-speed steel's microstructure.

7.3.1 TYPES OF CARBIDES IN HIGH-SPEED STEELS

In terms of carbide type, the same discussion as for cold work tool steels, in Section 5.3.2.1, applies for high-speed steels. However, because high-speed steels have higher amounts of Mo, W, and V, the main undissolved carbides will normally be of the type M_6C (Mo- or W-rich) and MC (V- or Nb-rich), but not the Cr-rich M_7C_3 carbides seen in cold work steels (for a volume comparison, see Figure 3.12 in Chapter 3). The main advantage of the different types of carbides in high-speed steels is the higher hardness: around 1500 HV in M_6C and 2200 HV in MC carbides compared to 1600 HV in M_7C_3. Other than this difference, the behavior of carbides will be similar, constituting the main microstructural feature determining the abrasion resistance. A few comments are just added on both M_6C and MC carbides in the following paragraphs.

M_6C carbides are observed in practically all high-speed steels, emerging when the amount of Mo or W, combined with C, is above the solubility in austenite. For steels with 1%C, this will occur if W and Mo contents are higher than ~3%, meaning

---

* In all conditions, the tempering of high-speed steels is always carried out in high temperature, which enables the transformation of retained austenite. But, the same caution with retained austenite discussed for cold work steels (see Sections 5.3.1.2 and 5.3.1.3) have to be applied in high-speed steels: avoiding excessive hardening temperatures and also avoiding long times between hardening and tempering.

that amount of W or Mo above this level will be present as undissolved carbides in the final microstructure. In high-W grades, the M_6C-type carbides are formed directly from the liquid by eutectic reactions. In high-Mo grades such as M2, the as-cast eutectic morphology is based on M_2C carbides, with a feather-like morphology, as shown in Figure 7.6a. Those M_2C eutectics are metastable and are converted to M_6C and MC carbides during reheating for hot-forging or hot-rolling; Figure 7.6b

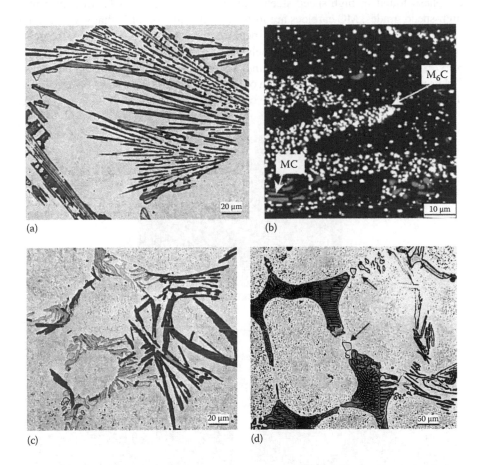

(a)

(b)

(c)

(d)

FIGURE 7.6 Examples of as-cast microstructures in M2 tool steels, showing (a) the feather-like morphology of M_2C carbides and (b) the transformation of those feathers, after reheating to forging or rolling, into M_6C and small amounts of MC. In (c), both M_2C (dark) and MC (light) eutectics are shown, after special etching with $KMnO_4$. And in (d), the primary MC carbides are pointed by the arrows; those carbides are larger and of cuboid morphology because they precipitate in the liquid phase and tend to grow in all directions following the most stable crystallographic planes, rather than getting elongated when formed by an eutectic reaction. Some of those MC carbides are also shown in (b). (Images from (a), (c), and (d) are from light microscopy and obtained from Barkalow, R.H. et al., *Metal. Trans.*, 3, 919, 1972.) The image in (b) is a backscattered image from scanning electron microscopy, where the phases with heavier elements become lighter, which is the case of the M_6C carbides rich in Mo and W.

shows M_2C feathers that have been "broken down" into M_6C and MC carbides. This change in M_2C contributes to the more refined morphology of the carbides in the microstructure, which is important for toughness and grindability. More details of both eutectic solidification structures can be found extensively in the literature, such as in References 8 and 9.

MC carbides, also explained in Section 3.3.3.1 (Figure 3.32), constitute the hardest phase found in high-speed steels and tool steels as a whole. In commercial high-speed grades, MC carbides are formed by the amount of V that was above the solubility limit, which quantitatively speaking means above ~1%V for ~1%C high-speed steels. The use of high-speed steels with very high amounts of V, such as above 3%V, is not common in commercial high-speed steels unless they are produced by PM routes. The reason is the formation of large primary MC carbides that precipitate directly from the liquid and not in the eutectic microstructure.* Both morphologies

M2 (0.9%C 5%Mo 6%W 2%V): 64–65 HRC

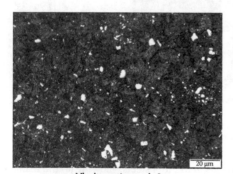

Nb alternative grade I
(1.1%C 3%Mo 3%W 1%V 1.7%Nb): 64–65 HRC

Nb alternative grade II
(1%C 2%Mo 2%W 1%V 1.2%Nb 1%Si): 63–64 HRC

FIGURE 7.7 Examples of niobium-alloyed high-speed steels recently developed and patented. Deep-etched samples, in the transverse direction of 30 mm round bars. (From Mesquita, R.A. and Barbosa, C.A., Hard alloys with lean composition, Patent: US 8,168,009, Priority: 2006; Mesquita, R.A. and Barbosa, C.A., High-speed steel for saw blades, Patent: US2009/0123322, Priority: 2006.)

---

* In the literature, those carbides are often named particle-type, primary, or blocky carbides, while the carbides formed by eutectic reactions are named eutectic carbides.

are shown in Figure 7.6c and d. While the refinement of carbides is always interesting for adhesive wear and toughness, the grindability aspect is of special interest in high-V, high-speed steels. Because of the high hardness, large MC carbides are not easily removed during the grinding operations for manufacturing of the tools and may cause local overheating on the tools' cutting edge, which in turn leads to reduction in hardness and even cracks. When large and concentrated MC carbides are present, these grinding issues become worse. Therefore, it is important to reduce the size of those primary MC carbides, which basically depends on the amount of C, V, and N, as described in Section 3.3.3.1 and in Reference 10.

Another important aspect of high-speed steel carbides is the potential of using niobium as the primary source for MC carbides. As explained in Chapters 3 and 5 (see Figure 3.10 and Table 5.2), Nb is a very strong carbide-forming element and the hardness of Nb-rich MC is higher than that of V-rich MC, leading to additional contribution to the wear resistance of high-speed steels. In addition, the low solubility of NbC in austenite leads to a large volume fraction of carbides without the need for the addition of large amount of niobium. In other words, while roughly 3%W, 3%Mo, or 1%V is needed to pass the solubility limit in 1%C tool steel, Nb will form undissolved carbides for amounts as low as 0.1% (see discussion in Figure 3.12 and Section 5.3.2.1). One concern in the literature is related to the size of the primary carbides, but this is often caused by over-addition of Nb, disregarding the characteristic that it practically does not go into solution and so smaller amounts are necessary to form primary carbides. In addition, the low levels of N, used to control the V-rich MC carbides, also apply for Nb-rich MC, as described in Reference [11]. So, applying those adjustments, Nb carbides may be an opportunity in high-speed steels, as shown by the examples in Figure 7.7.

7.3.2 DISTRIBUTION OF CARBIDES IN THE MICROSTRUCTURE OF HIGH-SPEED STEELS

In terms of carbide distribution, the mechanism in Section 3.3.3.1 shows that the carbides are formed during solidification, by the enriched liquid in the interdendrite spacing. And, during rolling, this carbide network structure is broken and aligned. This evolution may be observed in Figure 7.8. In this figure, the microstructure is classified according to its degree of homogeneity, A being the best and C the worst, according to SEP 1615 standard. It is easy to see that the as-cast structure with concentrated carbides is directly related to small bars with also heterogeneous distribution, or, in other words, the deformation in hot rolling is only able to elongate the heterogeneous areas but not eliminate them. So, the classification C of SEP 1615 shows, for smaller bars, thick alignments or stringers of carbides. This type of microstructure is considered of inferior quality, because concentrations of carbides are the preferred routes for crack propagation and also affect tool distortion during hardening. Although C and other alloying elements may be optimized, it is important to remember that the coarser or finer distribution of the final carbides will primarily depend upon the solidification (ingot geometry and casting variables) and the amount of deformation during rolling (see discussion in Section 3.3.3.2).

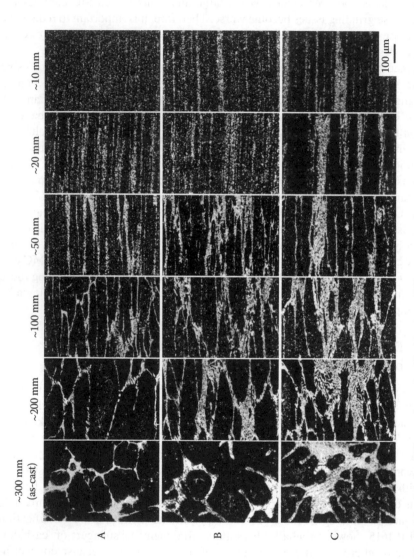

FIGURE 7.8 Carbide distribution for different sizes, classified in three different levels, from A (better) to C (worse), according to SEP 1615. (Adapted from Stahl-Eisen Prüfblatt 1615, Microscopic and macroscopic test for the image ordered carbide distribution of high-speed steels, January 1975.)

While those coarse morphologies are easily identified in large bars (dimensions above 100 mm/4 in.), the same observation is not trivial in small-sized bars, with 10 mm or less. Small sizes are also the most important for actual cutting tools, which usually are made in diameters less than 12 mm (½ in.). The problem of analyzing carbide distribution in those small diameters lies in the high scattering of results, which in turn relates to the large deformation in the forming process. Because the ingot dimensions are much larger than the actual bars, one sample in a hot-rolled coil may refer to a good (not segregated) area of the ingot while another sample, a few meters ahead, may show a completely different distribution of carbides, being related to another area of the ingot. So, the microstructural evaluation of small-diameters high-speed steel bars should be done not by one sample but by several random samples taken from the coil or bar ends. This is actually exemplified in Figure 7.9, which presents a statistical evaluation of several samples in lots produced under different processing conditions. With this statistical evaluation, the differences in the process can be visualized, when one distribution is shifted to better classification than another. Another suggestion to indirectly evaluate the final distribution is to take samples and observe the billets, which can be evaluated over a bigger area by macro or micro tests. Charts like Figure 7.8 may be used to qualify the billets and then anticipate how the quality in the final rolled product will be.

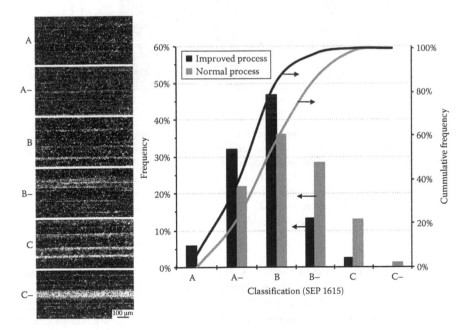

FIGURE 7.9 Examples of statistical evaluation in small diameter (~8 mm or 5/16 in.) high-speed steels. The classification follows SEP 1615, but with intermediate classifications between the A, B, and C original from SEP. Carbide distribution improves from C to A. About 100 samples for each condition were evaluated.

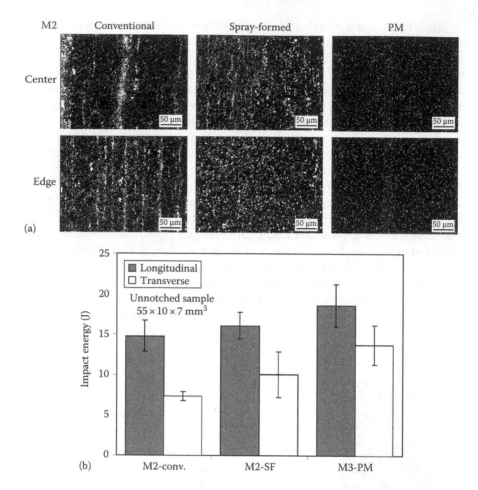

(a)

(b)

FIGURE 7.10 Variation of the microstructure and mechanical properties of M2 high-speed steel cast with three different methods: conventional casting, spray-forming, and powder metallurgy. (a) Carbides in longitudinal direction. (b) Toughness. (Adapted from Schneider, R. et al., The performance of spray-formed tool steels in comparison to conventional rout material, *Proceedings of the Sixth International Tooling Conference—The Use of Tool Steels: Experience and Research*, Bergstrom, J., Fredriksson, G., Johansson, M., Kotik, O., and Thuvander, F., eds., pp. 1111–1124, Karlstad, Sweden, 2002.)

To conclude this section on carbide distribution, a few comments on alternative casting methods are included, since the distribution of carbides may be dramatically improved by the refinement of the as-cast structure. As explained in detail in Chapters 3 and 5, the solidification with faster cooling rates leads to smaller as-cast eutectic microstructures, which will then have better carbide distribution in the final bars. This effect is actually shown in Figure 7.10a, which compares the conventional casting to spray-formed and PM cast material, showing the advantage of microstructure refinement and toughness properties (Figure 7.10b).

For simple tools, like twist drills, the cost of PM grades may be too high for large commercial applications. But, in other larger or high-demanding tools, the use of PM grades is economically viable due to the higher homogeneity and consequent better performance. More discussion on this topic is given in the application of PM grades in Section 7.4. The spray-forming process, although known in the literature, does not constitute a well-established technique in the industrial field and no regular supply of high-speed steels produced by this technique is found in the market.

7.4 PROPERTY COMPARISON AND SELECTION OF HIGH-SPEED STEELS

The same approach done for the other tool steels in the previous chapters is also applied here for high-speed steels, with all properties compared in a schematic diagram, as shown in Figure 7.11. This diagram shows the main properties with respect to performance but also the facility to grind (or grindability of) high-speed steel, because this is the main operation during tool manufacture. Five main groups of high-speed steels may be identified: (1) compositions for the most common applications, such as M2 or T1; (2) steels with higher amounts of V or Co, leading to higher hardness and wear resistance, such as the M3 class 2 grade (referred to as M3:2); (3) the so-called super-high-speed steels, with large amounts of Co to increase the tempering response; (4) PM grades, some grades with Co to improve the temper hardness; and, lastly, (5) semi-high-speed steels such as M50.

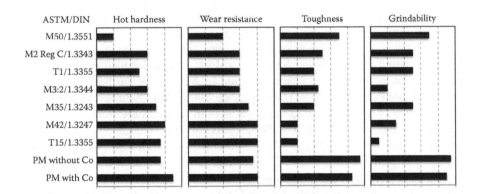

FIGURE 7.11 Qualitative comparison of important properties of the main tool steel (the larger the bar, the higher the property). *Notes:* PM grades are usually produced by steel companies in nonstandardized compositions and in the present discussion they are just split in two groups: alloyed with Co and nonalloyed. (Chart build with data of ÖNORM EN ISO 4957:1999, *Tool Steels*, Publisher and printing: Österreichisches Normungsinstitut, 1020 Wien; ASTM A600-92, Standard specification for tool steel high speed, ASTM International, Materials Park, OH; Erasteel, Grade datasheets for high speed steel products, Available online at: http://www.erasteel.com/content/asp-steels-0, Accessed April 2016.)

7.4.1 M2 AND OTHER HIGH-SPEED STEELS FOR STANDARD TOOLS

M2 is by far the most common high-speed steel used in industry because of the balance of properties such as hardness after tempering, tempering resistance, wear resistance, toughness, and grindability. It is commonly said in industry that M2 represents more than 80% of the volume of high-speed steels produced and used for standard cutting tools like twist drills. Regardless on the specific quantity produced, M2 is the main reference in high-speed steels for cutting tools. Most literature and also much of the data in this chapter refer to M2. Being so versatile, M2 is also used in several other applications, such as in some cold work tools and even in hot work conditions.

Similar to M2, T1 is another grade that can be classified as general-purpose material in standard tools. It was the first high-speed steel composition, developed in the beginning of 1900s by Crucible Steel Co., but with large contributions from the work of Taylor and White on the heat treatment of high-C, high-W alloys. Interestingly, the focus of Taylor and White was on reducing the manufacturing time by improving the cutting speeds, which was limited by the heating and consequent softening of the low-alloy, high-carbon steels used prior to the availability of high-speed steels. T1 then emerged as a perfect solution, with higher resistance to softening (tempering resistance). T1 was then the main tool steel for machining tools until World War II, when it was substituted by Mo-type high-speed steels (M-type) because of the shortage of W in the international market.*

The exchange of Mo and W as alloying elements in high-speed steels is based on their similar chemical behavior of these elements, both forming the same undissolved carbides in the final tools microstructure and the same tempering carbides. In fact, Mo tends to form M_2C eutectics during solidification, which is metastable and converts to M_6C as explained in Section 7.3.2; but W forms directly the M_6C eutectics. This makes the hot-forming of M-type grades usually easier, with the natural breakdown of the eutectics (Figure 7.6b). Independent of those small differences, W-based (T grades) and Mo-based (M grades) grades tend to show similar performance as long as the volume amount of Mo and W carbides is the same. Since the atomic mass of W is roughly twice that of Mo, there is a rule of substitution where 1Mo = 2W, and in many cases both elements are treated together as equivalent tungsten, or $W_{eq} = W + 2Mo$. If this equation is applied to M2 and T1, it is realized that both have W_{eq} of ~17%. The same happens for M1, with 8%Mo and 1.5%W, but also with W_{eq} ~17%. This relation is then useful to compare different steels, for example, to analyze some current trends of substituting M2 by grades with higher W, which has been done in China because of the large reserves of W in that country.

---

* During World War II, Germany and other countries involved in the war were the major suppliers of W to the world. North America, however, had reserves in Mo, and extensive research was done to develop M-type grades, and today M2 is the most common grade based on both elements actually: that is, 5%Mo and 6%W.

7.4.2 Modified M2 for Higher Performance: M3 class 2 and M35

Small modifications in M2 may be applied to improve the cutting performance, and this is done in grades M3 class 2 (or M3:2)* and M35. Both are classified as special high-speed steels, or HSS-E.

The main modification in M3:2 is the addition of 1%V more, which roughly doubles the volume fraction of MC carbides (since about 1% of the V is in solid solution, increase from 2%V to 3%V in total content leads to undissolved V of 1%–2%). This change in the microstructure increases the wear resistance but not necessarily the tempering resistance (hot hardness). So, M3:2 high-speed steels are mainly used in fine cutting tools, such as taps, but not in normal tools as a mean of increasing the cutting speed.

M35, however, has added Co for increasing temper hardness of M2. This enables higher hardness, contributing to wear resistance, but also intensifies the tempering response and increases the hot hardness. Therefore, higher cutting speeds may also be achieved. The advantage of M35 is that the carbide distribution, in terms of undissolved carbides, is very similar to that of M2, thus not affecting the grindability as the V effect in M3:2. M35 is used for special tools and also some large tools (above 20 mm diameter).

7.4.3 Super-High-Speed Steels

Large amounts of Co, usually in the range of 8%, may be used to increase the tempering response and achieve hardness levels above 67 HRC. These additions may also be followed by higher amounts of Mo, in the case of M42, or very high V, in T15. Both lead to a combination of high hardness and high amount of carbides, which increases the wear resistance and also the applicable cutting steep. Because of the high V and low grindability of T15, M42 is more common in several machining tools, especially those like milling cutters and high-speed drills. Within tool manufacturers, M42 grade is known as HSS-Co.

7.4.4 Powder Metallurgy High-Speed Steels

High-speed steels produced via PM solve the main problems of conventional steels such as the following:

- Improve the carbide distribution and homogeneity of properties, such as toughness and wear resistance.
- Enable the production of very high alloy composition, which would not be possible by conventional casting because of the large amount of carbides and the consequent low hot ductility during hot-rolling.
- Reduce grinding problems (by refining the carbides), enabling the increase of the amount of MC carbides by higher contents of V and Nb.
- Reduce warping and other heat-treatment distortions during heat treatment.

---

* M3:1 is similar to M3:2, but has lower carbon and is therefore less common in industry.

So, the performance of PM high-speed steels is well above that of normal steels, either in cutting or hot-forming tools. Nevertheless, the higher cost of those materials limits their application to high-performance tools, in which the cost benefit is usually in favor of PM materials.

In terms of composition, traditional PM grades are PM23 (similar to M3:2), PM30 (M3:2 with 8%Co), and PMT15 (traditional T15 grade, but with refined MC carbides by the PM process). They usually achieve high hardness even after hardening at temperatures lower than traditional grades, because of the easier dissolution of their fine carbides. And, as in traditional high-speed steels, compositions with high Co lead to higher temper hardness and superior performance.

However, the most interesting compositions of PM high-speed steels are those developed by the steel manufactures but not standardized. Taking advantage of the very wide range of compositions that may be cast using the PM process, new balance of properties may be achieved, usually aiming at high hardness (above 66 HRC) and high amount of carbides, mainly of MC type.

7.4.5 SEMI-HIGH-SPEED STEELS

Although the earlier discussion was concerned with increasing hardness and amount of carbides to achieve better cutting performance of high-speed steels, some applications are not so demanding. Examples are tools for hobby and nonindustrial use, such as drills for manual drilling, and also handsaws or other sawing machines for soft materials. For those applications, the use of high-speed steels with lower amounts of alloying elements—although affecting the properties—may lead to advantages in terms of reduction in the steel cost.

Steels falling in this category are often named *semi-high-speed steels*, the main examples being ASTM M50 and M52, as well as DIN 1.3333. They achieve hardness in the range 62–64 HRC, but not a minimum 64 HRC as M2. The tempering resistance is also smaller (see tempering diagrams, Figure 7.3b), which means that hardness will drop faster with the heating, limiting the cutting speeds. The low content of alloying elements also reduces the amount of undissolved carbide, so wear resistance is reduced. Therefore, they are limited to the above-described nonindustrial applications, which are the typical for semi-high-speed steels, as well as some cold work tools or mechanical components.

REFERENCES

1. E.M. Trent, P.K. Wright. *Metal Cutting*, 4th edn. Boston, MA: Butterworth-Heinemann, 2000, pp. 132–249.
2. ÖNORM EN ISO 4957:1999. *Tool Steels*. Publisher and printing: Österreichisches Normungsinstitut, 1020 Wien, 1999.
3. ASTM A600-92. Standard specification for tool steel high speed. Materials Park, OH: ASTM International 1992.
4. Erasteel. Grade datasheets for high speed steel products. Available online at: http://www.erasteel.com/content/asp-steels-0. Accessed April 2016.
5. Böhler Edelstahl. Technical catalogue for tool steels. Available online at: http://www.bohler-edelstahl.com/english/1853_ENG_HTML.php. Accessed April 2015.

6. Villares Metals. Technical catalogue for tool steels. Available online at: http://www. villaresmetals.com.br/pt/Produtos/Acos-Ferramenta. Accessed April 2015.
7. R.A. Mesquita, C.A. Barbosa, C.S. Gonçalves, A.L. Slaviero. Effect of hardening conditions on the mechanical properties of high speed steels. *International Heat Treatment & Surface Engineering*, 5, 36–40, 2011.
8. M. Boccalini, H. Goldenstein. Solidification of high speed steels. *International Materials Reviews*, 46(2), 92–115, 2001.
9. R.H. Barkalow, R.W. Kraft, I. Goldstein. Solidification of M2 high speed steel. *Metallurgical Transactions*, 3, 919–926, 1972.
10. H. Mizuno, K. Sudoh, T. Yanagisawa. The influence of alloying elements on the morphology of MC primary carbide precipitation in Mo-type high speed tool steel. *Denki Seiko*, 55(4), 225–236, 1984.
11. R.A. Mesquita, C.A. Barbosa. Hard alloys with lean composition. Patent: US 8,168,009, Priority: 2006.
12. R.A. Mesquita, C.A. Barbosa. High-speed steel for saw blades. Patent: US2009/0123322, Priority: 2006.
13. Stahl-Eisen Prüfblatt 1615, Microscopic and macroscopic test for the image ordered carbide distribution of high speed steels. January 1975 (original in German: Mikroskopische und makroskopische Prüfung von Schnellarbeitsstählen auf ihre Carbidverteilung mit Bildreihen), Januar 1975.
14. R. Schneider, A. Schulz, C. Bertrand, A. Kulmburg, A. Oldewurte, V. Uhlenwinkel, D. Viale. The performance of spray-formed tool steels in comparison to conventional route material. *Proceedings of the Sixth International Tooling Conference—The Use of Tool Steels: Experience and Research*, Eds. J. Bergstrom, G. Fredriksson, M. Johansson, O. Kotik, F. Thuvander, Karlstad, Sweden, 2002, pp. 1111–1124.

Index

Milton Keynes UK
Ingram Content Group UK Ltd.
UKHW040106071024
449327UK00019B/854

9 780367 782573